Shaking Up the World

Other Books by James D Paulk Jr.

Swimming for Our Lives

Shaking Up the World

Stories of the Naval Academy Class of 1957

Compiled by James D Paulk Jr.

BOOKLOGIX
Alpharetta, Georgia

The authors have tried to recreate events, locations, and conversations from their memories of them. The authors have made every effort to give credit to the source of any images, quotes, or other material contained within and obtain permissions when feasible.

Copyright © 2024 Naval Academy Class of 1957
Individual chapters are copyrighted to the respective authors. Compiled by James D Paulk Jr.

All rights reserved. No part of this book may be reproduced or transmitted in any form or by any means, electronic or mechanical, including photocopying, recording, or any information storage and retrieval system, without permission in writing from the authors.

ISBN: 978-1-6653-0673-7 - Paperback
ISBN: 978-1-6653-0674-4 - Hardcover
eISBN: 978-1-6653-0678-2 - eBook

These ISBNs are the property of BookLogix for the express purpose of sales and distribution of this title. The content of this book is the property of the copyright holders only. BookLogix does not hold any ownership of the content of this book and is not liable in any way for the materials contained within. The views and opinions expressed in this book are the property of the Authors/Copyright holders, and do not necessarily reflect those of BookLogix.

Library of Congress Control Number: 2024906758

∞This paper meets the requirements of ANSI/NISO Z39.48-1992 (Permanence of Paper)

Cover photo: Second Lieutenant Zach Placek is leaving Bancroft Hall after graduation, as we did, to begin his new career. The photo view of him leaving via the Rotunda was taken by Carolyn Andros from the steps leading to Memorial Hall. Cover photo supplied by Carolyn Andros. Photos in sections supplied by authors.

073024

RON

Friend, Classmate, Naval Aviator, Rear Admiral, US Navy, President Naval War College, Superintendent US Naval Academy, President US Naval Academy Alumni Association, Founder of Naval Academy Foundation, Founder of the Class of 1957 Chair in Naval Heritage for the History Department, Distinguished Naval Academy Graduate.

Contents

Foreword	xi
Introduction	xiii
US Naval Academy Class of 1957 History	1
Hijacked	6
Busted by an Admiral	18
Gregory Peck	20
Doing a Treaty with the Russians	22
1954 Army-Navy Game	26
Auschwitz Survivor, Art Aronson	28
USS *Clamagore* Makes a Port Visit To Bridgeport, Connecticut—July 4, 1971	36
Colonel's Revolt	40
The Welder	42
The Critical Failure That Led to GPS	44
Taking Care of Business	50
Starting a Tradition	53
Twenty-Seven Memorable Years in the NFL	57
Sneezing Crickets	64
Sneak Attack on a Junk	66
Secret Shenanigans	69
Sharks in the Pacific	73
Shot Down over Enemy Territory	75
Building a Submarine	87
Counterinsurgency and Survival Training	92
Eisenhower and Nixon Experience	95

A Greek Holiday	98
Charlie Duke, Astronaut	103
Engineer to Newspaper Owner	111
A Soviet Black Sea Cruise	114
Finding Thresher	119
Glasgow Marine Court	121
Guardian Angels	124
How I Spent My Childhood in a War Zone	126
KGB Encounters	136
Locked in the Jail!	141
Dependent's Cruise	145
Fritz, a Marine in Vietnam	149
Get off That Submarine	155
Gunship History in Laos	158
Highline Adventure	167
Left on the Bridge	169
Message in a Bottle	172
Invitation to Hell	174
A Navy Junior's Far-East Odyssey, 1938–1940	184
Long Carrier Deployment	187
My Career Was over before It Started	192
Pearl Harbor, the Beginning of an Odyssey	195
Positive Leadership	201
Reporting Aboard	203
Pearl Harbor Attack and Rescue at Sea	206
Possum in the Church	209
Bob Nevin, Hydronaut	212
A Navigation Nightmare in the Georgia Straights	219
Bumping *Queen Mary* from Drydock	221
Marine Officer Escapades	223
Naming a Submarine	227

National Reconnaissance Program	229
Pearl Harbor Attacked	236
Ship Stability or Lack Thereof	238
St. Mary's Football Coach	241
Submarines Can Be Targets Too	242
Tales From a Fish, Chicken, and Corn Farmer	244
The Farewell Ball Dress	248
Worms	251
Underground War Story	254
Women in the Navy	256
The World War I Fourteen-Inch Naval Railway Gun	259
Flag Lieutenant Follies (1)	262
Flag Lieutenant Follies (2)	265
Fueling in Bergen, Norway	267
Helicopter Down	269
Insubordination	275
Janitor to Admiral	277
Submarine Ice Exercise	280
The Great and Grand Tear Gas Mission	283
The London Garden Party	287
The Safe	289
The Stolen Wallet Drill	292
Woo Poos Tried to Kill My Wife	294
Taking Command	297
Softball Game	300
1964 Thresher Search	302
Fire Drill	304
A Rescue at Sea	307
Acknowledgments	*317*
Gallery	*319*

Foreword

Charlie Hall

ONCE JIM HAD published his first book, *Swimming for Our Lives,* and saw how well it was received, he began to listen to folks who said he should do another book. He thought a bit and decided to open his horizons beyond his adventures to those of his friends and colleagues from the Naval Academy Class of 1957. While this is a far different sort of book than *Swimming*, it has taken on a life of its own. Many of Jim's classmates pitched in to help with all the aspects of publication. Some wrote of their own adventures, some helped with researching stories for accuracy, and some assisted with editing and proofreading stories written by others. All in all, the project evolved to a class-level undertaking that has led to this collection.

These stories range far and wide from activities at home and abroad to military missions, combat, and otherwise. The result is a grand view of the Class of 1957 in all its various roles and undertakings. This class has made many differences in this world! Find those differences in the book! Look for an astronaut, an NFL referee, many technology experts, Admirals and Generals, and some just plain Sailors, Marines, and Airmen going about their businesses. Businesses? Well, much of those were keeping the Nation and the world safe for you and me to live fine lives. Some were improving the quality of life for all of us. Others were bearing arms in combat to preserve democracies here and abroad. The Class of 1957 made its mark all over the world, in many disciplines, military and civil, and many walks of life.

Read all about them and their exploits, their joys and sorrows, their wins and a few losses. And, in the end, you will surely see that the Naval Academy Class of 1957 has indeed *Shaken the World*—and it will never be the same. I am proud to be a member of the Class of 1957, a friend of Jim's, and a contributor to this wonderful effort.

Introduction

IN JUNE 1953, young men all over the United States and a few foreign countries learned their dream had come true—they had an appointment to the US Naval Academy in Annapolis, Maryland. At the end of the month, there we were in stately Bancroft Hall—our new home for the next four years. We did not have much of an idea of what we were to eventually face, but that first night was all about meeting as many of our new classmates as we could. It was so exciting, maybe the most exciting day of my life. My new colleagues all seemed to be valedictorians, presidents of their high school class, or captains of their football teams. Since I was none of those things and was so impressed with my classmates, I began to wonder why I was there and how I might fit in. Shortly thereafter, I realized my two years at a military college had provided me with maturity and a head start on academics. Both of those elements were important to my eventual success.

Without them, I doubt if I would have made it to graduation. Hundreds were there that first night who did not.

Well, here we are, approaching ninety years of age, and I'm still impressed—in awe really—of my classmates and specifically, with their accomplishments, adventures, humorous incidents, and major contributions to the world we live in. Thus, *Shaking Up the World*, is my way of sharing stories for you to enjoy about the lives of people I care so much about. I had a classmate who walked on the moon! Now, isn't that a remarkable achievement to come from such a small group of men? There is that and more. Classmate George Bouvet put it best in his summary, "US Naval Academy Class of 1957 History," the country got its money's worth. Yes, taxpayers paid for our education, but these stories about our lifetime adventures, humorous and otherwise, and the impressive professional achievements of these men support George's words.

In the process of writing this book, I thoroughly enjoyed learning things about my friends I had not known and laughed out loud at some of the stories. But most of all, I enjoyed getting to know my classmates better. I believe you will have the same experience.

—Jim Paulk

US NAVAL ACADEMY CLASS OF 1957 HISTORY
(A Good Return on Taxpayers' Money)
George Bouvet

WHAT HAPPENED TO MY CLASSMATES from the Naval Academy over the years?

On arrival at the Academy for Plebe Summer in 1953, 1,160 Midshipmen were admitted. Of these, 848 graduated in 1957 with 847 receiving a BS degree in general engineering, and almost all received commissions into the Navy, Air Force, or Marine Corps. (The Air Force Academy had not graduated their first class yet, and West Point and the Naval Academy provided a source of regular officers for the Air Force.) During the four years, 312 dropped out voluntarily or for academic, disciplinary, or medical reasons.

About 67 percent of the class went into the Navy (160 went to Flight School), 24 percent went into the Air Force, and about 7 percent went into the Marine Corps. The choice of the specific assignment was based on a "number" drawn at random for each person.

When we entered the Academy, the minimum obligation was three years of service in the Air Force and four years in the Navy or Marine Corps. Those who selected Naval Air or Flight School were obligated for an additional eighteen months of service. Submariners had another year of obligated service.

Many of my classmates intended to make the service their career, but many were unsure about a twenty-year and thirty-year career commitment immediately after graduation. However, 68 percent did stay in the service for a minimum of twenty years.

We graduated when Eisenhower was president, and the Cold War was in full force. Russia launched its Sputnik later that year, and the missile era was beginning.

The first nuclear submarine had just been launched, and Admiral Rickover was leading the "cutting edge" of the nuclear Navy.

Several classmates were lost during the Vietnam War, having been shot down while on missions for the Navy or Air Force. Also, many died in flight training or during accidents flying off carriers. President Reagan stated in a letter to our class in November 1986, "I pause with you as you remember the forty-eight graduates who gallantly gave their lives while on active duty."

Twenty-five became admirals or generals. Two were prisoners in Vietnam for over five years after being shot down. One became an astronaut and walked on the moon. Another became a hydronaut and went down over eleven thousand feet to examine the debris field of the submarine, *Scorpion*. Another led the design, advocacy, and development of the now-famous GPS (Global Positioning System).

Many received advanced degrees in engineering fields or management (many from MIT, Stanford, or Harvard). Some went on to get PhDs. Others left the service and entered the business world, getting MBAs and joining the ranks of corporations. Others started businesses.

Several became lawyers. A few entered the medical field, some became politicians, and a few ministers or priests.

It would not be possible to cover the careers of all my classmates here, but let me try to summarize the highlights of a few:

Bob Brown, Lieutenant Colonel USAF, deceased. Bob entered the Air Force and became a pilot. He flew 299 combat missions in F-100s over Vietnam and won four Distinguished Flying Crosses, 16 Air Medals, and the Vietnamese Gallantry Cross. In 1972, Bob was again assigned combat missions, flying the new F-111 from Thailand to North Vietnam and was killed in action while flying a mission on November 7, 1972. He was posthumously promoted to Lieutenant Colonel.

Leo Hyatt, Captain US Navy, retired. Leo went into Navy Air and in 1967, while flying his thirty-third mission to Vietnam, was shot down. He spent five years and seven months in captivity, tortured after suffering extreme injuries when ejecting from his plane at 850 miles per hour. His captivity was like that of John McCain, who graduated one year after our class. Leo received a Silver Star for his heroism while in captivity. After being released, he returned to flying status, commanded a Skyhawk fighter squadron, and retired after twenty-eight years in the Navy.

Charlie Duke, Brigadier General USAF, retired. Charlie took a commission in the USAF at graduation and went to Flight School. In 1962, the Air Force sent Charlie to MIT for a master's degree in aero engineering, and after graduating, he went to test pilot training at Edwards Air Force Base. Charlie then became an astronaut in the Apollo program. He was Communicator on Apollo 11, the first lunar landing, and was on the backup crew for the Apollo 13 crew. He went on to fly Apollo 16 and spent seventy-two hours on the lunar surface. Since retiring, Charlie has been active in Christian ministry work.

Brad Parkinson, Colonel USAF, retired. Brad joined the USAF and later was sent to MIT for a master's degree and then to Stanford for a PhD in aerospace and astronautics. He was then assigned to the Air Force Test Pilot's School as an instructor. While Deputy Head of the Astronautics Department at the Air Force Academy, he spent most of the year in Southeast Asia, where he flew 150 hours of combat missions working with the digital Fire Control System on a gunship. Brad subsequently led a Department of Defense team in developing the first GPS and was awarded the Draper Prize in engineering for his leadership role. He is presently a professor emeritus at Stanford and was also the Program Manager for NASA Gravity Probe B (which was a test of Einstein's General Theory of Relativity).

Bruce DeMars, Admiral USN, retired. Bruce went into diesel-powered submarines early in his career and into nuclear-powered submarines afterward with extended tours of duty away from his family. He served as Admiral Rickover's successor and head of Submarine Warfare. Bruce left active duty in 1996, after forty-three years in uniform, and served on several corporate boards.

Earle Smith, Captain USN, retired. Earle was a Battalion commander as a Midshipman and captain of the Navy football team, where he was an All-American end. Earle entered the Navy and became a Submariner. His career almost ended when a torpedo shot turned back into his submarine at a depth of two hundred feet. Eighteen men were trapped, and Earle—equipped with an oxygen mask—entered the toxic-filled compartment and rescued the men. He was awarded the Navy-Marine Corps medal for his leadership, heroism, and devotion to duty.

Fritz Warren, Lieutenant Colonel USMC, retired. Fritz was the Midshipman Brigade Sub-Commander at the Naval Academy and Class President for thirty-seven years. He served as an enlisted Marine before entering the Academy and was commissioned Second Lieutenant upon graduation. When the Marines landed in Vietnam in 1968, Fritz requested a front-line combat assignment, which was granted. As a Major, Fritz was forced to take command of a Marine detachment at Dai Do, when his two senior officers were killed and another five hundred were wounded. For that, he was awarded the Legion of Merit medal.

Not mentioned above are many classmates who helped win the Cold War, serving on ships and submarines for extended periods, flying strategic bombers or tactical fighters, or spending many years of duty in underground ICBM sites.

After leaving the Service, 266 in our class owned at least one business, 169 classmates were Presidents or VPs of companies, and 333 served as CEOs. At age seventy, 38 percent were working full- or part-time.

It is my privilege to have gone to the Naval Academy and to be associated with those mentioned above and with my other distinguished classmates.

I think our class represents a good return on taxpayers' money, don't you?

[*Compiler*: George Bouvet was a Guided Missile Launch Officer in the USAF and later became Boeing's first 747 Marketing Director.]

HIJACKED
Jerry Barczak

JUNE 14 to July 1, 1985, was a once-in-a-lifetime event for me. I became a hostage of the then-current Mideast terrorists. After being safely home for over twenty years later, I still have not researched the incident, read books or magazine and newspaper articles, or viewed TV program tapes I'd been given. All are boxed and put away for any of my descendants who might be interested in granddad or great granddad.

Over the years, the conversation regarding the incident was brought up by someone else, as I normally did not initiate it. Other than a few questions in public, I have entertained only one reporter's scheduled interview and have made one safety video for my employer. The following is my recollection of events in which I participated or that which I learned during discussions with others in captivity.

On graduation, I was commissioned into the Civil Engineering Corps (CEC), so my career was in the maintenance and construction management of military facilities throughout the world. I retired from the Navy in 1980 and started my second career in civil engineering positions in the Middle East. In June 1985, I was working on constructing an Egyptian air base and took leave to attend my daughter's high school graduation from Mission Bay High School in San Diego.

I left Cairo with a planned connection in Athens, Greece, to the United States, but my plane was late, and I missed the connection. As I was walking in the Athens airport, evaluating my travel problem, an airline hostess recognized my concern and asked about my situation. On hearing my needs, she said there was a TWA plane going to Rome, where I could get a direct flight to California. Fortunately, it was leaving shortly so she called the plane to wait. It had left the parking area, so they took me out to the tarmac in a truck with steps, and I climbed into the plane. What luck! I was the last person on TWA flight 847, with seat 4D in the first-class section, and my luggage could catch up with me in San Diego.

My row mate in seat 4C was a young man from the Chicago area who was a reserve Army Major on active duty to assist in the planning of a joint-military training exercise. Although I didn't know it at the time, the passengers included US Navy-enlisted men who had been working on construction projects in Greece. They carried passports indicating they were active-duty military. Also, I believe some passengers carried diplomatic passports.

Within fifteen minutes, we took off, and while the plane was climbing, a passenger ran past our row to the front of the airplane. My row partner said, "He's got a grenade." All was out of my sight, and I was taken completely by surprise. My initial reaction was, "Urh, what faction?" because there had been several aircraft hijackings in the early eighties. All passengers were caught off guard, but I believe Seaman Stethem and one other Navy man rushed forward to stop the hijack. The hijacker stopped them with a handgun and told them to return to their seats. Later, I found out there were two hijackers on the flight and a third was denied passage and left behind in Athens, where he was arrested. During the flight, air hostess Uli Derickson was able to converse with the hijackers in German and she was excellent in passing any word very quietly, advising the passengers of what was happening but did so very carefully as the hijackers were against it.

The next announcement from the plane captain was that we had a couple of passengers who wanted to go to Beirut, Lebanon. We were told to sit quietly in our seats. After a short while, those of us in first class were moved to the rear of the plane; and I was ordered to sit on the floor in the small space between the seat row and the fuselage in one of the last rows. Everyone was ordered not to talk, to bend over, and to place our heads between our knees. For me, being on the floor in such a small space was very uncomfortable and later I developed leg cramps. (Later, this led to an individual event for me.) The hijackers collected all ties and belts from passengers so these could not be used against them.

We landed in Beirut, where some children and women—believed to be Muslims—were allowed to leave the plane. All the rest remained on board while the plane was refueled. After about two hours, we took off for an unknown city, and were told to remain in our bent-over positions. At this point, I now had a seat.

It turned out we landed in Algiers, where the hijackers originally wanted to go but fuel was too low to fly there, hence the stop in Beirut. In Algiers, more children and women were allowed to leave, but here also, the beatings of an Army Major and two US Navy men began. The beatings were done in the first-class area so very few—possibly only the plane staff—observed these actions. The men were told to crouch in the doorway and were then pounded with a seat arm the hijackers had broken from a seat. We heard the blows and the cries of the beaten.

In Algiers, the hijackers apparently did not get any satisfaction, and after about five hours when the plane was refueled, we took off again. During the flight, we learned we would land again in Beirut. The pilot planned to land even though the Beirut tower said, "No," and would not turn on the runway lights. Essentially, he said, "Here we go," and proceeded to descend as best he could on his instruments. Later, I found out the hijackers wanted to trade the aircraft and all passengers for hundreds of Lebanese or Muslim prisoners held by Israel. We were political hostages.

Day two started in Beirut. It is here that Seaman Stethem is beaten again and executed. When this occurred, the flight attendant, Miss Derickson, announced, "You may want to plug your ears. You will soon hear a sound you might not have heard before." Then there was a sharp pop—a gunshot. What it meant, we did not know at the time, but found out later that Seaman Stethem was killed and thrown off the plane to the tarmac. Early in this stop, more gunmen came on and it was the Hezbollah, the group to which the hijackers belonged, and about twelve Amal militia. At this stop, I think the military, diplomatic, and persons with Jewish-sounding names were taken from the plane.

No one had been allowed out of their seats, even to go to the toilet. Eventually, after a rather long time, one person raised her hand and asked permission to go. After some conversation between the hijackers in their language, they said, "Okay, but leave the door open." One of them then stood by the open door. After the first person had gone, another raised her hand and was okayed. After this happened four or five times, the guards were frustrated and changed the routine. Now anyone could go, one at a time, after the previous one was back in her seat. Without a doubt, we all were physically relieved after a while.

After a very short stay, and with the addition of more gunmen than the original two, we took off not knowing where we were headed again. It turned out we returned to Algiers and stayed for the rest of day and night two. Here, the potential third hijacker, arrested in Athens, arrived under the care of Greek officials, one of whom was very distinguished and personable. He told the hijackers that, for their potential third hijacker to be released back to them, he wanted all the women, children eighteen and younger, and any Greek persons to stand. As different individuals stood and said a few Greek words, he said, "Okay, outside," then he chose a few older, over eighteen, but young-looking children to be freed. As a closer, he insisted people with health problems should also be released. In the end, most of those on the plane were released at this stop, which left about thirty of us as plane hostages, plus those taken off earlier to captivity in Beirut. On day three, we flew back to Beirut and within the next two days, a few more Greek citizens and medical-problem persons were released. The hostage count was then thirty-seven male passengers plus three male pilots. Also, later, two or three more men who had extreme health problems were released from captivity.

It was robbery day. In the previous days, the hijackers occasionally stole such items as watches, earrings, or other jewelry from the passengers, but now each of the remaining hostages was called individually to the first-class section, where they and their wallets were searched. Cash, traveler's checks, and jewelry were taken. In my case, while going through my wallet, they passed over my retired military identification card, probably not realizing the color differed from the active-duty cards. Returning to my seat, I decided to remove my ID card and put it in my suit coat breast pocket, which later brought it back to their attention.

During the robbery phase, they rifled through everyone's carry-on items, and I lost the souvenirs and gifts I was bringing home. Also, on this day, they broke into the plane's luggage area and went through many of the suitcases, scattering much of the luggage in the area near the plane while it was parked well off the active area of the runway.

Everyone lost their traveler's checks and one of the men lost three very expensive cameras. He was an architect and lost the reviewed engineering drawings of palaces he was designing for some Saudi Arabians. Luckily, I think he had the originals in his home office in the States. Later, the hijackers initiated what I thought must be some reaction to their guilt—an act to replace or trade for what they had taken. They passed out traveler's checks to most of us without regard to names. I never found mine, but I found the owner of those given to me and passed them back.

Regarding the expensive cameras, the one who took them gave the architect back a low-cost 35mm camera.

Since most of the initial passengers were left off in Algiers, all hostages were male. While we were awake during the day, we were seated in the center section of the plane, three in each row on each side of the passageway. At night for sleeping, they allowed us to split up and each take a three-seat section for a bed. Food and drink were provided from the kitchens of the airfields wherever we were. The additional guards that had come onto the plane were much less cruel than the original two, yet it was obvious they were part of the operation. They had their weapons and had also placed what they said was an explosive system on the deck of the fuselage.

During the day, as we sat in our required area, their leader would talk with us to improve his English. In my opinion, he was an excellent leader and had complete control of his men. If he had been on the right side, I think he would have done well. When asked, he told us he had been trained in Egypt, and that he belonged to the Islamic Jihad. That was the first time I heard mention of this group. So we were being held by the Hezbollah, Amal, and Islamic Jihad, with the Amal probably being the most stable of the three. Another interesting fact was that the parents and families of the hijackers knew these men were in radical organizations and had hijacked the airplane.

In the early morning of day four, at approximately two or three a.m., we were awakened a few at a time by the guards and told to leave the plane. Only the pilots remained. When told to leave, I said I had to get my jacket from the seating area I was in during the day, but they said, "No, move out." Later, I learned the reason we left the plane was because they feared a commando-type attack on the plane.

We were all taken to an apartment house and split into two or three groups. A day or two later, eight of us were moved to a house which was our stay until release. Luckily, after we left the plane to go to the apartment or house, we were able to shower, shave, and somewhat wash our underclothes. They gave us toiletries from the airline stock. In each place, there were a few beds, but most of us slept on the floor throughout the space available. The stay in these areas was rather boring and time moved slowly. In fact, throughout the entire time of our captivity, I felt rather at ease, expecting to be released eventually, but I also realized the guards would shoot any of us at any time if ordered to do so by the leaders. (Note: I didn't keep notes, so the happenings of the following days are not identified to the day but are essentially in somewhat chronological order.)

There are three principal memories for me of this time. The first happened on an airplane. One of the guards had a habit of walking the aisle and pressing the trigger on his handgun, a forty-five. He walked by me, pointed the gun at my head, and pulled the trigger. Then he laughed. It sounded like a loud snap and happened so quickly I just froze. Thank goodness I heard the laugh.

The second happened in the apartment when I suffered very bad leg cramps, which I thought were the result of the cramped aircraft position and maybe some dehydration. At night, I tried to sleep against a wall with my legs raised up on the wall, hoping for some relief. One of the guards was a fellow named Ali. I believe he and my fellow hostages told one of the terrorist leaders of my trouble, and they took me to a doctor who prescribed a relaxer prescription to be picked up at a drugstore. Three men took me to the doctor and pharmacy in a Mercedes Benz and sped through the streets as if no other cars were on the road. I returned to the group with the medicine, which did help stop the leg cramping. I admit, while away from the group, I was not sure I would be returning.

The third incident had to do with my military ID card. Since my suit coat was left on the plane, they found the ID card when the pockets were checked by the guards. At this time, I was with the other seven in the house and one of the captors came and asked if I was in the Navy and what I did. I told him I retired in

1980, was a civil engineer, and was working on the construction of water and sewer lines, roads, and buildings in Egypt. He wanted to know specifically what I did in the Navy, so I told him I built and maintained military facilities, such as airbases or ship ports.

His English was not very good, and he appeared confused by my answers, so he left. Later, another man talked with me and understood everything, especially my retirement and that I had not been active in the Navy for the last five years, so he told me to just forget the incident. His English was completely American, so I asked him how he came to speak so well. Apparently, he had attended college in California. What a surprise! The Hezbollah was extremely angry with the USS *New Jersey's* shelling of the coast with their sixteen-inch guns and one of the Hezbollah guards had lost all members of his family.

Overall, the group of Hezbollah, Amal, and Islamic Jihad had very strong ideas. This was obvious when they were extremely excited about the arrival of the media persons to Beirut. I overhead a few speaking in English and they were amazed a full plane load of newspaper and TV people came to cover their hijacking and holding of hostages. One said something like, "Wow, a whole plane full. We'll sure get our views to the people of America and the world." They intended to get the world behind their desire to have Israel release the hundreds of Muslim prisoners being held and to embarrass the USA.

Now looking back, some incidents were humorous. While in the first apartment, Ali, the guard, took two or three hostages at a time upstairs to another apartment to have tea and cookies with his mother and family members. Once, he set his rifle in the corner of the room, forgot, and left it, only to later come back excited and ask where it was. Since it was obviously no advantage to any of us at the time, it was pointed out to him, and he retrieved it immediately. He looked extremely relieved.

Another happening took place in the house where there were eight of us. The house had an outside second-floor landing, which was always occupied by one of the armed guards—a young man, about eighteen or nineteen, with the habit of cocking his pistol and pulling the trigger while pointing the barrel downward. One day, our food was delivered by a couple of men in a van while the young man was on guard. He did his usual trigger-pulling habit, but this time a shell was in the chamber, and he shot the van's driver in the shoulder. He was taken away immediately but returned a few days later. We asked him about the shooting, and he said he was disciplined and ordered to visit the hospital and stay with the injured man for a few hours every day.

While in the house, the eight of us got along well, but we all kept relatively to ourselves. We had not known each other before, and each had his own worries. We passed the time by playing cards that were found in the house, blackjack, poker, hearts, and crazy eights. Fortunately, we were allowed to listen to the radio, so we heard news about us a couple of times a day as the status of our hijacking was always being reported by the BBC. It was one day in late June when there was almost no news, so we concluded that something was up. But what?

Twice we were taken out to very nice hotels for dinner, but mainly to be presented to the newspaper and TV reporters. Many of the thirty-seven were interviewed, but one fellow and I kept to ourselves. Now and then I would walk in the background of someone being interviewed, hoping I would be on camera so my family could see I was all right. As it turned out, someone must have kept track of the interviews because the two of us were eventually told we had to sit for a short interview by the media. I've blanked out whatever was asked and how I answered.

After the news lessened on BBC, we felt our government must be negotiating, and maybe we would be released on the Fourth of July as a grand gesture. On June 28, we were taken to a hotel for a big dinner affair, one with a large cake. During dinner, we were told we would be going home but no date was given. Obviously, our hopes were raised. The next day, we were taken to a schoolyard under heavy guard, more than any previous time. The media was there also. We expected to be freed that day, but not all hostages were there; a few were still in confinement somewhere. I talked to some of the Navy men, and was told they were being held in a basement prison. Apparently, the negotiations underway did not meet the captors' satisfaction and, later, I learned there were internal differences among the three factions holding us. We were sent back to our holding areas very disappointed.

On June 30, in the late afternoon, we were back at the schoolyard with all the hostages. The Red Cross was there with a convoy of automobiles into which we were loaded, told we were being released, and to be driven to Damascus, Syria. Though the distance was about one hundred miles, the trip was about five hours—a long five hours because of the many checkpoints and our psychological anxiety.

We arrived in Damascus very late at night and entered the Sheraton hotel to be met by the US ambassador who told us we would not be staying at the hotel. Rather, we would be taken directly to a waiting US Air Force C-141, which we boarded and took off in the dark of the night for Rhein-Main Air Base in Frankfurt, Germany. At this point, I believe we all thought we were on our way home. It was early morning on July 1, 1985.

At the time of the hijacking, my brother, Edward, was a Chief Petty Officer aboard a carrier in the Mediterranean, working in the Intelligence Division. When the ship learned I was a hostage, he was immediately removed from intelligence and reassigned to another division. He took leave when he heard we were being released and met me at the Frankfurt hospital, where all of us were taken for a health check-up and a debriefing interrogation. When I saw Edward, I was very pleasantly surprised. It was great to have a family member nearby. He stayed with me until I returned to my home, and then to Memphis, Tennessee. I also visited my other family members in San Diego and West Allis, Wisconsin.

While in the hospital, we were given physical and psychological checks. I thought I had weathered the incident fine, but I guess

I fooled myself. When I read my record, I apparently did have some psychological downs. The doctor had written that I looked older than my age and that I answered easy questions slowly or incorrectly. While in Frankfurt, I was asked numerous times for interviews or to schedule appearances on TV after I returned to the US. But I decided not to. As it turned out, my employer was happy I never voiced for whom I worked, which at the time was General Dynamics Services Company. As for in-depth newspaper interviews, after I returned, I did agree to one with a *Wall Street Journal* reporter my son had established a good relationship with.

The interrogation, or debriefing, was held very soon after we were in Frankfurt. Each of us was taken individually and asked to just talk about our remembrances. Then we were shown photographs, asked to identify any we recognized, and sign our name on the back of those photos. In my case, I mentioned the house across the street from the apartment, what it looked like, and that there were military-type vehicles in the garage. I drew a sketch of the front of the building showing the under-building garage entrance from the street. Also, I gave the medicine packaging and bag from the pharmacy to a second set of interrogators after I was with my family in Wisconsin. These had names and addresses of interest.

TWA flew us freed hostages and our family members from Frankfurt back to the United States on July 2, a most pleasant trip. We landed at Andrews Air Base in Washington, DC. Just prior to my departure from the plane, someone gave me a small US flag which I waved as I came through the aircraft door. That is one picture I remember seeing in the newspaper.

Even today, I have two very fond memories of my return that day. First, as I was stepping through the plane doorway, I saw classmates and posters of the USNA Class of 1957. The second was the honor of stepping off the plane and disembarking to shake hands with President Reagan and Mrs. Nancy Reagan, whom I will always remember.

As to my family discovering my situation. When my scheduled

flight that was to land in San Diego did not arrive and there was a reported plane hijacking, my son felt something was up. He checked with the airline and was told to phone a certain number. During that call, he was told I was on the hijacked plane, and was asked to provide contact names and numbers. And of course, shortly thereafter the media had the names.

With my home in Memphis and my family in three places, my oldest son took charge because he remembered I had talked with all my children before starting work in the Middle East back in 1980, telling them that if anything ever happened, they should keep my background history to themselves and not talk about my work. With that in mind, and since the newspapers and TV had enough other people who would talk with them, time passed with minimal pressure from the media.

I did lose my carry-on baggage, watch, passport, money, military ID card, other valuables, and suit coat, but I did not lose my Naval Academy 1957 ring since I did not carry it with me to my overseas assignments. My checked luggage showed up in San Diego as ticketed and I retrieved it after my return to the States. My daughter's graduation classmates wore yellow flowers on their graduation gowns at Mission Bay High School, and she was asked to say a few words. I did have a few welcome home parties, and, after a month, I returned to my work in Egypt.

It's been a long time and I do have two concerns when I compare the terrorists of today to the captors of TWA Flight 847. First, today's terrorists are much more violent and ferocious, and it is likely many of the thirty-seven of us would probably have been killed. Second, the terrorists of today live within communities with good Muslim people, yet these good people do not report the bad in their midst, which is difficult for me to understand. Luckily, at the time of my incident, the radicals' purpose was to bring world attention to their programs, and captured aircraft with passengers were normally returned. From that time on, every day has been great for me!

BUSTED BY AN ADMIRAL
Carol Marryott, Wife of Ron Marryott

Rear Admiral Ron Marryott
Superintendent of the US Naval Academy.

THIS RECOLLECTION is from 1986, a week before Christmas. Ron, Superintendent of the Naval Academy, and I were enjoying some downtime as most of the classes and activities were over for the semester. This evening, we invited the Mayor of Annapolis, Dennis Callahan, and his wife, Brenda, to join us for dinner in town. The Callahans were a delightful couple and always good company. Also at our party were Mark and Mary Tempestilli. At that time, Mark was Ron's Flag Lieutenant, and he and Mary had become very dear friends. They were lively and fun to be with, a great addition to our evening.

We gathered at Buchanan House, our quarters, which incidentally had recently been renamed Farragut House. As soon as our little group assembled, we left for the short walk into town. Leaving by the front entrance facing Blake Road, we headed toward Gate Three. Ron and Dennis were in the lead, and as we neared the Naval Academy Chapel, we could see and hear two people fighting, yelling, and rolling around in the grass, almost knocking over the Nativity scene there.

Our two leaders reached the combatants first and quickly pulled the two up by the neck of their jackets. There in front of us were two very embarrassed and shocked Midshipmen. Not only were they caught, but their captors were the Superintendent of the Naval Academy and the Mayor of Annapolis.

I could not hear what was said, and I cannot tell you what the penalties were, but I am sure this is a Christmas memory that will never be forgotten by those present at this surprising brawl.

GREGORY PECK
Earl Piper

IN JUNE OF 1947, 402 Coast Boulevard, La Jolla, California, was the summer getaway of a famous Hollywood actor. It was a small single-story cottage with a cozy front porch, and that house is still there today. It overlooks a quiet beach just a few steps away, a place where the soft sound of ocean surf resonates all day. At age eleven, I spent many hours hiking the hills of La Jolla with my buddies, but my favorite pastime was swimming at the Cove or body surfing at Windandsea Beach.

One of my best friends at the time was Pete Andrews. He was a wiry, skinny, blond-headed kid; he came from a troubled home and never had rules to follow. He was related somehow to a famous Hollywood actor and called him "Uncle." One fine day, Pete and I were walking along Coast Boulevard in our bathing suits, headed for the beach at Casa De Mañana.

Pete said, "Let's stop and get something to eat at Uncle Greg's house." Pete was always hungry. We went up to the door of this small cottage and walked right in without a knock. Uncle Greg and Aunt Greta were there and welcomed us in.

They gave each of us a sandwich as I recall, and asked us where we'd left our shoes—but of course, we never wore shoes to the beach. After our short visit, Pete and I proceeded down to the Casa, and he told me his uncle's name was Gregory Peck. That afternoon, when I got home, I told my mother what had happened. She seemed excited about our visit with Gregory Peck and said he was a well-known actor who was born in La Jolla and had gone to the same elementary school I was attending at the time. I was excited too because I had shaken hands with Mr. Peck and his wife.

About a week later, I happened to be downtown in La Jolla village at the old hobby shop. I was getting some balsa wood for my airplane modeling hobby. On the way home, I stopped at the pharmacy store in the main entrance of the Colonial Hotel, where I wanted to buy a candy bar but didn't have enough money left after my balsa wood purchase. A man was standing next to me who was just checking out. It was Gregory Peck. I recognized him—as had everyone else in the store—and he recognized me as well, Pete's buddy. Somehow, he knew I'd been eyeing the candy. He reached into his pocket and gave me a fifty-cent piece. A half-dollar! At age eleven, that was two weeks' allowance! I stuttered out, "Gee, thanks, Mr. Peck." Then we went our separate ways.

When I got home, I told my mom, but I don't think she believed me . . . at least not until I showed her my fifty-cents worth of candy bars. I'd spent it all! In later years, my mother would tell this story much more often than I did. The first film starring Gregory Peck that I ever saw was *Twelve O'clock High* in 1952. I loved it. He was a good actor. Whenever I see his movies now and recall our brief meetings in the past, it always brings an odd feeling of kinship . . . albeit fantasy and totally distant.

DOING A TREATY WITH THE RUSSIANS
Ted Kramer

I'M DELIGHTED and somewhat surprised someone suggested my experiences of treaty negotiations with the Russians for inclusion in the book, as that was, indeed, a memorable and unique tour of duty for this Surface Warfare Officer.

The one thing I remember most from the experience was to expunge any semblance of editorial pride you may have. Going through the murder boards of an "article," "agreed statement," "common understanding," or any other treaty-related substance you may have drafted for the Delegation was worse than any Academy engineering work I ever encountered.

A great portion of the treaty work was dedicated to nothing more mundane than language conformity. There are no prepositions in the Russian language so we had to be very careful about how our own prepositions would be interpreted, e.g., heavy bombers "equipped *for*" connotes a bomber that is *capable* of being equipped with whatever and could be counted in the launcher aggregate if and when it is equipped; as opposed to "equipped *with*" which connotes a bomber already equipped with whatever and therefore already counts in the current aggregate.

Translation differences also played a role, e.g., our "post-boost vehicle" translated to "self-contained dispensing mechanism" in Russian, and some of our words, e.g., "capability," had completely different meanings in the Russian language. Verb tense played a major role: "*The Soviet Union will not equip* heavy bombers with . . ." is fine, except it doesn't prohibit the Russians from moving their heavy bombers to Poland or any other country to have them equipped with whatever. Hence, we preferred, "*Heavy bombers of the Soviet Union will not be equipped . . .*" That sort of language play blocked several treaty loopholes.

But so much for the didactics.

Socially, dealing with the Soviets could be engaging. During a cocktail party at our ambassador's home in Geneva, Switzerland, my wife accidentally dropped a canapé on the Ambassador's expensive carpet. Noting her embarrassment, one of the Soviet ministers quickly came over and told her, "Don't worry, Mrs. Kramer, just tell the Ambassador the Russians did it."

Or they could be boorish. During a garden party at the Soviet Mission celebrating their November Revolution, one of my counterparts, Colonel Grigoriev, who was the stereotypical mirthless, doctrinaire Soviet, came over to make small talk with me. I noticed that the fly in his pants was wide open, so I maneuvered him close to the Soviet Ambassador and loudly proclaimed to the Colonel that his fly was open. This didn't seem to bother him at all. He just unobtrusively zipped himself up in front of the Ambassador and his wife and just as loudly complained the new "western" suit he bought in Geneva that week had a zippered fly, which was not nearly as practical as the buttons on the fly of a Russian-made suit. Hence, western technology will never replace Soviet traditionalism and will eventually lead to our demise.

But most of their diplomats had a good sense of humor. We were arguing with their Minister Smolin one day, trying to get our definition of cruise missile range into the treaty. Our somewhat dubious definition was something called "maximum effective range," which only measured the straight-line distance from the launch point to the target no matter what meandering course the missile would take. The Soviets, of course, desired the definition of "maximum range" of the cruise missile to be entered into the treaty, that is, the actual distance the cruise missile would travel from the launch point to the target, including the distance it traveled throughout its entire route.

Smolin, as an example, questioned that if he would drive from Geneva to Divonne, a French casino town fifteen kilometers over the Swiss border, by way of Paris three hundred kilometers away, would only the fifteen kilometers from Geneva to Divonne count on his car, despite the long distance he traveled? We argued ingenuously that yes, we would only count the fifteen kilometers because what was important was getting from Geneva to Divonne, no matter what route you took. Disgusted, an exasperated Smolin shouted, "Never mind! The French wouldn't give me a visa anyway."

In line with technology, we were always warned about Soviet spying, and their crude attempts to see what we were going to present or what our position was before we met with them. They rented an apartment in a building a block away from our building and, indeed, our technicians discovered the tell-tale signs of electronic snooping coming from the building. The Soviets were using long-range, high-powered cameras to take pictures of the papers on our desks.

Accordingly, I had to keep my blinds permanently closed to thwart the Soviet cameras. This negated a marvelous view of the Geneva Botanic Gardens across the way from my office and, on a clear day, a spectacular view of Mont Blanc seventy kilometers away. There were also stories that the Soviets had cameras hidden in the light fixtures over the conference table at their Delegation site so they could see what notes Ambassadors Earle or Warnke had on their papers. The Ambassadors, of course, were aware of this and took precautionary measures to hide what they were reading.

The most blatant surveillance attempt, however, came in the Soviet conference room. We always sat on the side of the table facing the entrance with our backs toward the windows overlooking the grounds of the compound. There was a huge curtain that was always shut to keep the light out of the room. One day, a loud whirring noise came from behind the curtain and a roll of tape from a tape recorder started spewing out from underneath the curtain. The Soviets used old-style reel tape recorders and one of the reels got stuck and the tape started spewing out. We all got a laugh out of it, mostly at the expense of the Soviets who made a valiant but futile effort to ignore what was happening. Believe it or not, that's the way it was working with our Russian friends.

1954 ARMY-NAVY GAME
Earle Smith

HAVING BEEN A PRISONER OF WAR for seven years in Vietnam and shot down during his time as an A-6 pilot, Captain Jack Fellows, USN, retired, got his choice of duty when he was released. He requested and was selected to serve as the Football Representative for the US Naval Academy.

He told us that those in the "Hanoi Hilton" (prisoners, that is) used to each have a night to tell a story — they would often review a film, book, event, etc. Jack chose the 1954 Army-Navy game (Navy 27–Army 20) where the Navy was No. 1 on defense in the nation and the Army was No. 1 on offense.

There was a controversy about whether I caught a pass in the end zone or if I dropped it (of course, I caught it). Jack said, "I said that story every time my turn came around for seven years, so, if you dropped the ball, I don't want to know."

My wife, Sandra, was quoted in the "Philadelphia Daily News" in an article for the 100th anniversary of the Army-Navy game saying, "It amazes me. When we go back for class reunions or

football reunions, we see Admirals, Generals, and guys still on active duty. They can't remember the important stuff, like their wife's birthday, but they can remember everything about the football games they played in."
   GOD BLESS AMERICA—
   BEAT ARMY!

AUSCHWITZ SURVIVOR, ART ARONSON
Sam Coulbourn

Art Aronson on a summer cruise, 1956.

ARTHUR ARONSON was indeed an interesting figure in the Naval Academy Class of 1957.

I met Arthur Aronson sometime that summer when we arrived to start our Plebe year. He asked me to join him as a roommate, along with Milt Bank. So, when it was time for the Brigade to return for the academic year, we moved to our Plebe-year room in the twenty-first company, Sixth Battalion, which resided on one floor of the Sixth Wing of Bancroft Hall, the largest dormitory in the world.

I soon discovered Arthur was a foreigner—he was a Jew, born in Tomaszow-Mazowiecki, about fifty-five miles southwest of Warsaw, Poland, where his father had owned and operated a textile factory. The family had been herded by the Nazis into the concentration camp at Auschwitz (Oswieczim), Poland.

He told us of the day he and his family were herded onto a box car in his hometown along with many others. As they arrived at the prison, there was dark smoke blowing over the tracks, and it was the smell of burning flesh. Art later learned the smoke was from huge incinerators.

Although he was only about eight years old, Art was given the tattoo on his left arm given to all Auschwitz prisoners. His tattoo was B-1204. Think about that!

Art got separated from his family. His dad went to a work camp and his mother and sister must have died soon after arriving in Auschwitz. The Nazis considered Jewish women useless. They killed them lest they reproduce more Jews. But abled-bodied men were kept to labor, and I expect they figured Art would soon be old enough to work. His father was sent to a separate area and worked in a factory making supplies for the Nazi war effort. He was one of several men building ammunition boxes for the Wehrmacht. Aronson, the crafty elder, invented a way to build these wooden boxes so they looked fine when they were inspected, but when loaded with ammunition and in use in the field, the bottoms would fall out.

The Polish Jews also sabotaged socks that had been knitted in the camp for shipment to Nazi soldiers on the Eastern front by quietly slashing through each box of socks, so that the socks were useless for protecting their wearers in the brutal winter.

Art and his father survived, and after the camp was liberated, Arthur, aged ten, and separated from his father, made his way to Łódź, Poland. His father went to New York City. The family's long-time governess, Maria Sroka, a gentile, found Arthur after being freed and looked after him.

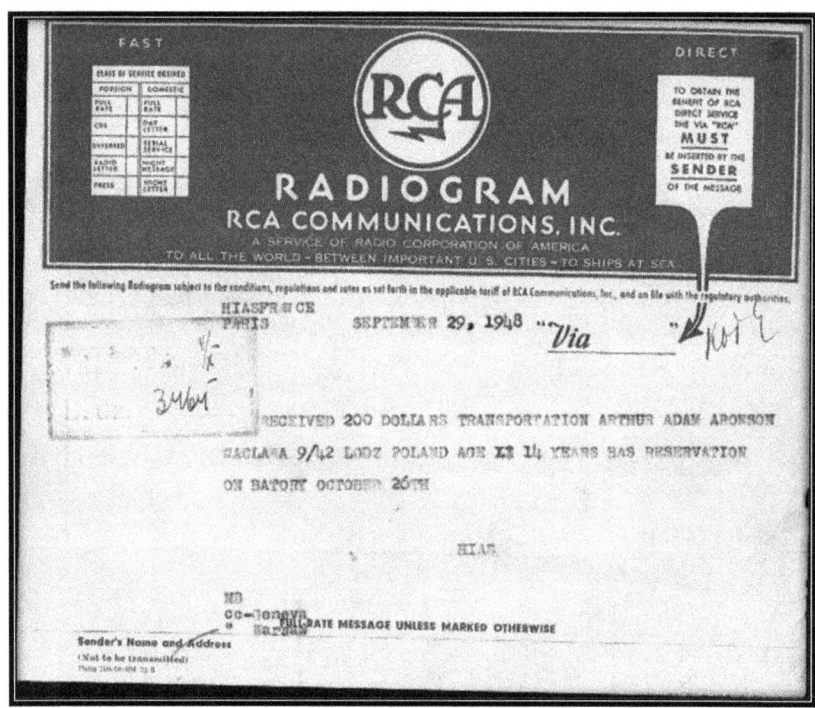

HIAS notice that fourteen-year-old Arthur was sailing for America aboard SS *Batory* in 1948.

The Hebrew Immigrant Aid Society, helping Jewish refugees in the US since 1907, connected Arthur with his father and arranged for him to be sent to New York, arriving in 1948. In about 1952, Arthur was sent to Houston, Texas, where he spent a year at Rice Institute. Someone there must have helped get him appointed to Annapolis.

Twenty-First Co. Midshipmen, 1956 L to R: Aronson, Coulbourn, Peterson, Missailidis, and Whiting.

Arthur's English was so good, I had not realized he was a Polish Jew for several days.

Arthur was probably the most brilliant person I have ever met. He spoke Polish and flawless English, German, Yiddish, and Russian. He could gobble up whatever subject the Academy threw at us—calculus, thermodynamics, steam engineering, naval gunnery, physics, English, history, navigation, and then took the time to help his less brilliant classmates understand.

Art was also an excellent artist and developed a cartoon style he used to illustrate Naval Academy magazines all four years of his time at Annapolis.

For spring leave in our second-class year, we went to see Art's New York stomping grounds so I could meet his father. He told us about the sabotage in the Auschwitz factory.

Aronson's artwork in The Log, ca. 1954.

In our blue Midshipman uniforms, Arthur and I visited the United Nations on this short spring vacation in New York. By this time, I also spoke a good bit of Russian, and we enjoyed using it to converse without others being able to understand. At the UN, we visited the only conference of delegates in session at the time, a round table of women observing International Women's Day, discussing women's international issues. We were both wearing headphones, which could be switched to hear the discussion in several languages. We chose Russian and listened to this rather haughty, self-important Soviet woman say, "In the Soviet Union, women have freedoms that women in other countries can only dream about."

When you are wearing headphones and talking with someone—unless you are careful—you speak loud, and so all of a sudden, the women delegates around their circular table were looking at these two American Midshipmen, speaking Russian.

One had just said to the other, *"Kakoi bol'shoi govno!"* or, "What a load of ___!"

We were asked to leave.

That same day we visited the *Today* Show (1955), where Dave Garroway was in charge and joined the crowd of onlookers when the cast did their outdoor stand-up. We behaved for that.

When it was time to head back to Annapolis, we boarded the train at Grand Central. On the train, we sat near two young women, and Art, who was always more skilled at this than I, started a conversation with them. One was reading her history textbook, and we began to discuss it and asked them where they were from, etc. I noted one girl's name in her book and later wrote to her.

Art and I got off the train in Baltimore, and the girls continued to Washington. They were traveling from Boston to Washington for a visit with a friend there.

I ended up marrying that girl and she and I shared the Navy adventure in America and four other countries. Martha Jane died in 2018. We were married for sixty-one years. I shall forever thank Art for that meeting!

Arthur Aronson by Virgin Cannon at Annapolis, 1954.

Aronson was an expert at meeting pretty girls. Here, Art and two girls pose in front of Tecumseh at the Academy, 1956.

Art developed a neurological problem during his first-class year and so he was discharged from the Navy and went to work at Carrier Corporation after graduation.

I was aboard a destroyer operating in the western Pacific when I heard Art was working on the island of Guam, selling Carrier air conditioners. When we visited Guam, we had a short visit.

Art married Emily, from Long Island, and they were transferred to Carrier's office in Athens, Greece, and then later to Las Palmas in the Canary Islands. They finally settled in Syracuse, New York, Carrier's headquarters. He died of cancer in 2008. Emily died in 2021.

L to R: Coulbourn, Bator, Aronson, and Shea, 2003 Rockport, MA.

We met with Art and two other twenty-first company mates and wives after we had all retired, in 2002 and 2003.

We always thought that perhaps, Art was doing more than selling air conditioners in those years in Greece and Spain.

Emily and Arthur, 2002.

USS CLAMAGORE (SS-343) MAKES A PORT VISIT TO BRIDGEPORT, CONNECTICUT—JULY 4, 1971
Peter Boyne

ONE OF THE UNIQUE highlights during my command tour on *Clamagore* was a trip to Bridgeport, Connecticut, in 1971 to participate in the Barnum Festival and to celebrate the Fourth of July with the good people of the most populous city in

Connecticut. The father of one of my junior officers was the Chief of Police in Bridgeport, and it was based on his efforts that a port visit was arranged. The ship would be open to visitors and a contingent of the crew would participate in the city's parade. July Fourth fell on a Sunday that year, which meshed with the ship's operations schedule. *Clamagore* could arrive on Saturday, hold an open house for visitors, participate in the parade on Sunday, and depart Monday morning for planned weekly ops.

Bridgeport is situated on the Pequonnock River on Long Island Sound, about seventy miles from New London via I-95. The wives and young children were driven to Bridgeport, but several of the older children were allowed to ride the boat with their fathers. Paul, his good friend Michael, and Cathy were my guests. *Clamagore* arrived with a power boat escort and was tied up in Bridgeport Harbor. The townspeople were excited to be able to visit a "submarine" and began lining up for tours. The *Bridgeport Telegram* reported, "more than two hundred visitors an hour were conducted through the sub and lines at the Union Square dock where the vessel tied up had a waiting period of more than two hours."

During this time frame, there was a radical left-wing militant organization, the Weathermen, that carried out a series of domestic terrorism activities, beginning in 1969 through the 1970s, which included bombings, jailbreaks, and riots. The FBI described the group as domestic terrorists with revolutionary positions. Rumor had it that there might be an attempt to disrupt the festivities. Forewarned is forearmed. In addition to the ship's security, the Chief of Police arranged for police coverage on the pier and the boat. Fortunately, all was calm, and visiting hours were extended to ensure all visitors were accommodated. The parade was great fun, with families cheering the troops from the bleacher seats.

Clamagore got underway Monday morning; fortunately, the tide was high. For a brief period, the boat sat on the bottom when the tide was low. The boat was headed to Earle, New Jersey, to exchange weapons. The Naval Weapons Station's distinguishing feature is a 2.9-mile pier in Sandy Hook Bay, where ammunition can be loaded and unloaded from warships at a safe distance from heavily populated areas. The route to New Jersey

continued down Long Island Sound, to the East River, into New York Harbor, and then on to the weapons station. For this
trip, a pilot was necessary to avoid all the commercial traffic, including the Staten Island ferry. *Clamagore* was on the surface, which provided New Yorkers along the river's edge an unusual sight. As we passed the United Nations building, almost every window was filled with people waving. Upon arrival in Earle, weapons were exchanged and *Clamagore* headed to the Long Island Sound op-areas for training.

These are good memories. Paul's friend Michael, to this day, still recalls the sunburn he got while topside as well as the fried chicken he had for lunch in the wardroom. Michael is retired now. He completed a career as a Captain with the Bridgeport Fire Department. Who could have known?

COLONEL'S REVOLT
Ron Goldstone

ONE OF THE MORE MEMORABLE perks of my tour of duty at NATO in Brussels, Belgium, was the opportunity to visit a NATO country once every three months. It was a family adventure. We pulled the kid out of school and away we went, by car to the host's designated city. During 1972–76 there were thirteen NATO members and we managed to hit all of them.

One trip took us to Istanbul. To get there, we had to cross Yugoslavia and Bulgaria. We were able to transport them but with a few anxious moments, since I was incognito, having been advised before my sanctioned travel to conceal my NATO identity.

Since I had previously visited Athens without the family, I promised my wife we would detour to that city on our way back to Belgium after we left Turkey. I assured her I had been to the Greek capital several times while deployed and she would enjoy seeing the Parthenon and other historic sites.

We arrived at the city's outskirts after sunset and navigated our way to our hotel. However, on the way, we encountered an eerie stillness about the city. No vehicular traffic; no pedestrians. The streets were deserted. I thought this situation was very odd. I recalled Athens as a very bustling, animated metropolis and so remarked to my somewhat skittish passengers that something was terribly wrong. This was confirmed shortly when we came upon a burnt-out bus and then a bank with its windows broken out. I immediately decided to change course and executed a quick 180-degree turn!

Upon turning the corner, we came face-to-face with a Sherman tank, its muzzle pointing directly at us. An armed Greek army officer waved us to the curb. He didn't speak English; I didn't speak Greek. But we each could communicate in halting French. He indicated he wanted us and our luggage out of the car. At this point, my wife and daughter were close to hysterics.

As mentioned above, I had been advised to conceal my NATO status, especially when traveling through Communist countries. But since Greece was a NATO ally, I threw discretion to the wind, whipped out my military ID, and feigned outrage at his demand. It worked. The officer saluted and sent us on our way.

We arrived at our hotel, parked the car, and with rifle shots resounding in the distance, raced into the lobby and safety. We soon learned we had unwittingly timed our Athen's visit coincidentally with the "Colonel's Revolt" which overthrew the civilian government. Was this considered a perk?

Surprisingly, all was quiet the next day, but all airports and ferry services were shut down. We took a leisure tour of the city, took in the sights, and, eventually, returned unscathed to Brussels.

Who would have thought that adventure could be recounted as a sea story?

THE WELDER
John Russell

THIS WAS A MEMORABLE EVENT for me that occurred during the second class year (junior) during one of our Electrical Engineering Lab sessions in the lower level of Sampson Hall.

By way of orientation for the reader, our EE courses included the study of AC and electrical currents, circuits, and other facets of electricity including hands-on time in a large, well-equipped laboratory. The lab in Sampson Hall was notorious among Midshipmen because it was a place where one mistake, or one distracted moment, could bring swift, shocking discomfort, sparks, smoke, and loud crackling sounds that garnered dark disapproval from faculty professors along with unwelcome laughter and ridicule from onlooking classmates. This was *the* lab that did not forgive lapses in concentration and is the setting for this story.

After the completion of a lab experiment, the established procedure was to first disconnect the power cables, which were inserted in the power trench in the floor, and then disassemble the experiment we had set up on the top of the lab bench, and lastly, to hang up the power cables on the racks located on the perimeter walls of the lab. Simple enough.

One dark and gloomy afternoon during the winter months, I was following all the foregoing established procedures and was proceeding to hang up the power cables. I was holding the wood handles of both cables in one hand and was walking across the room toward the racks, dragging the cables behind me. Suddenly I heard a loud explosion and turned to see a small mushroom cloud rising behind me. Apparently, the other ends of the two cables had somehow short-circuited as they were dragged across one of the power trenches. Not only were the other ends of the cables now welded together but the Academy's circuit breakers were tripped all the way back to the power station. Bancroft Hall, our dormitory, went dark—as did every classroom in the Yard (Campus). Classes for the remainder of the afternoon were canceled. Hooray!

That evening, during the evening meal, I was summoned by a loudspeaker to report to the Brigade Commander's podium. I was thereupon presented with a welder's mask as a token of the Brigade's appreciation. Loud applause and cheering!

Unfortunately, I was subsequently required to return the mask as it was technically US government property. I guess the Superintendent didn't share the Brigade's appreciation!

THE CRITICAL FAILURE THAT LED TO GPS
Brad Parkinson

IN A WORD, I HAD *FAILED*. My presentation, at the highest level of the Pentagon, to demonstrate a new, satellite-based navigation system was disapproved. The long table in front of my raised platform was populated with more Generals, Admirals, and Senior Civilians than I had ever addressed. I was the Program Director for US Air Force Project 621B and a very new Air Force Colonel. It was August 1973, and I had just briefed a system design I had inherited, requesting about $150 million for a full-scale demonstration. I didn't have time to contemplate my failure. Dr. Mal Currie chaired the meeting. As the Under Secretary of the Department of Defense, he controlled all of the military's research and development spending. Dr. Currie immediately asked me to join him, alone, in his third-floor Pentagon office.

About four months earlier, I had spent about two and a half hours with him, alone, in *my* tiny office at Los Angeles Air Force Station—an astonishing meeting, given the disparity in rank. I explained the tremendous value of a worldwide 3-D positioning system for the Department of Defense. With a Stanford PhD in guidance and control, I understood both design and technology intimately. The system design, indeed, could stand some refinement, but I had felt it would do for a demonstration.

Dr. Currie told me he strongly supported such a system, but he wanted an updated design that would satisfy the needs of all military services, suggesting I use the best technology I could find. He said he felt such a proposal could be approved.

We called that disapproval "Black Thursday," but in a way, it was Golden Thursday—it led to GPS.

I had been allowed to recruit a superb cadre of young Air Force Officer-Engineers. All had advanced degrees from outstanding universities. I resolved to call a redesign meeting, far away from potential opponents of change. We held that meeting in the Pentagon, over Labor Day weekend, 1973. The sole attendees were eight of my Officers-Engineers and two Aerospace Corporation Engineers.

Nine years earlier, Dr. Ivan Getting, the CEO of the Aerospace Corporation, advocated a new satellite-based navigation system and persuaded the US Air Force to fund a classified study of alternatives in 1964–1966. This concluded with a description of about a dozen different satellite-system designs and capabilities. The most difficult design required four satellites in the user's view and predicted worldwide twenty-four-seven, three-dimensional accuracies of about ten meters. It would use orbiting, hardened atomic clocks. This would become GPS.

A competing Naval Research Lab (NRL) concept, Timation, was also included in the earlier USAF study, six years before the NRL filed for a patent. The patent was finally issued to the Navy, in 1974. (Evidently, NRL was unaware of the earlier Secret USAF study.) Their patent described a two-dimensional system that required an atomic clock in each *user's* receiver. It was deemed too expensive and inadequate for general use, characteristics that ruled out the NRL concept in our redesign meeting.

Our weekend's effort was outlined in a seven-page Decision Coordinating Paper (DCP) that summarized the new proposal. Civil use would be enabled by promulgating the details of a "clear" signal. Thus, from its inception, we intended to make GPS available for civil use, but with no guarantee of availability.

I then made repeated briefing trips to persuade the decision-makers not to say no. By mid-December 1973, I received approval for a $150 million program to demonstrate GPS (our new name).

We went into a wartime development environment. We launched the first operational GPS satellite in forty-four months (February 1978). By 1979, all seven types of user equipment had been tested and demonstrated in eleven different vehicles and circumstances. GPS had proven every claim we had made for accuracy and coverage. In particular, it had demonstrated bomb delivery and military vehicle location accuracy that far exceeded anything in the Department of Defense inventory.

So, GPS was a much better "mousetrap," but the reaction of the USAF was extremely disappointing. Despite our testing success, the service did not want GPS. The USAF zeroed out the budget for GPS for a series of years in the early 1980s. Fortunately, civilian leadership in the Pentagon and the White House overruled and restored the funding.

Even with only the six-satellite test constellation, applications for time transfer to nanoseconds and Precision Land Surveys had begun by 1980. With twenty-four satellites, GPS was finally declared operational in December 1990. Demonstration of GPS value during the wars in Bosnia and Iraq completely reversed the views of the operational military. For many years, the Second Space Operational Squadron of the USSF has been a fully dedicated GPS operator and steward for both civilian and military users, worldwide.

Three events greatly accelerated GPS use. President Reagan guaranteed GPS to the world in 1983; civil use was no longer "at risk." In 2000, President Clinton officially abandoned any deliberate degradation of accuracy. During the same period, integrated digital circuits drove down costs and greatly increased capability. Today, a five-dollar GPS receiver can simultaneously receive over sixty channels of GPS as well as the European, Russian, and Chinese versions of Global Navigation Satellite Systems (GNSS). Typical accuracies are a few meters.

These three foundations assured civil GPS *availability*, with full *accuracy*, and with both the signals and receivers virtually *free*. They accelerated GPS adoption. Virtually, every element of the US critical infrastructure now depends on GPS. In 1978, I had forecast many of the applications, but the markets and manufacturers have far exceeded our dreams. The annual economic benefit has been estimated at well over a trillion dollars, without accounting for safety of life benefits. For example, farm tractors are now a $2 billion-per-year GPS market, with driverless control accuracies to a few inches.

However, the ubiquitous dependency on GPS has created concerns. In the UK, a study estimated, "The economic impact to the UK of a five-day disruption to GNSS was £5.2 billion."

Perhaps the greatest technical reason for concern is that the received GPS signal power is tiny: one-tenth of a millionth of a billionth of a watt (10^{-16} watts). This signal comes from forty-five-watt satellite transmitters located eleven thousand nautical miles away. Consequently, the weak GPS signal is potentially vulnerable to deliberate or inadvertent interference. Some users have sought augmentations by using lower altitude earth satellites, or ground-based radio navigation systems. Such efforts are understandable, but the US PNT Advisory Board has cautioned: "No current or foreseeable alternative to GNSS (Primarily GPS) can deliver equivalent accuracy (to millimeters, 3D), integrity, and worldwide, twenty-four-seven availability." They advocate protecting the signal and toughening the user's receivers to mitigate interference.

The single most effective toughening against GPS interference and false signals (spoofing) is the use of multi-element, Digital Antenna Arrays—DAAs—which create "nulls" in the direction of jammers to greatly reduce their effectiveness. In fact, in 1975, I anticipated the jamming issue and enlisted the Air Force's Avionics laboratory to develop a GPS receiver that would demonstrate all the receiver techniques to counter jamming (A/J) and spoofing. They partnered with Collins Radio, and, by 1978, they demonstrated a GPS receiver that could operate while flying directly over a 10-kW jammer.

Unfortunately, current State Department International Traffic in Arms Regulations (ITAR) preclude the use of DAAs with more than three antenna elements for nonmilitary applications. (The capability of DAAs greatly increases with seven or more elements). While well-intentioned, both the underlying design and technology of DAAs have been published extensively over the last fifty years, rendering this ITAR restriction ineffective in preventing the proliferation of this technology. For example, a Turkish Company (TUAL) currently offers a sixteen-element GPS antenna for any application or customer, worldwide. They claim it increases jam resistance by over one hundred thousand. With the advent of cheap digital components, arrays of up to ninety-one elements, with about a meter diameter, have been considered.

The notion that current ITAR restrictions prevent technology transfer, or that the US possesses a technological advantage over potential enemies is not factual. By disallowing our use of these technologies, the USG simply denies US civil aviation and other key users access to the well-known techniques. Unlimited DAA use could make these applications nearly immune to hostile or inadvertent interference. The restriction certainly does not restrict potential enemies. There is some evidence that the Russians were employing DAAs in the drones overflying Ukraine.

The world now has three other GPS-clone, worldwide systems. Civilians can freely use up to ten signals on four frequencies. Fifty or more navigation signals are typically in view. All these developments would benefit from DAAs. These foreign systems have a goal of surpassing the US GPS.

Despite any vulnerability, worldwide GPS applications continue to proliferate because of economic productivity, safety, and usefulness. Many billions of cell phones routinely provide users with locations to a few meters. Where are we headed with positioning navigation and time?

An example of the future is the use of GPS with other sensors to enable driverless long-haul trucking. There are over thirteen million truck drivers in the US. Long-haul driving is expensive, tedious, and potentially hazardous. Prototype autonomous trucks have already been demonstrated by three manufacturers.

The fiftieth anniversary of GPS's initial approval is a time for celebration, but also a time to prepare for the future. The most important step is for the US government to rescind all restrictions on Digital Antenna Arrays. These restrictions are ineffective in preventing the widespread proliferation of the best technology to toughen GPS against all forms of signal interference. Failure to do so will simply widen the gap, with other countries continuing to advance the well-understood state of the art and condemning US aircraft to remain susceptible to GPS jamming.

Perhaps the ultimate tribute to GPS is that knowledge of position is "taken for granted" and billions of people, worldwide, use the system each day. Engineers accept anonymity as part of our profession; such widespread use is the cherished payoff for us developers. But "taken for granted' will be misplaced if the US does not remove the fetters on our receiver industry.

Prince Charles Presents Brad Parkinson with the QE II Award for his GPS role.

TAKING CARE OF BUSINESS
Bob Fox

I ARRIVED IN WASHINGTON to be the Project Manager for Attack Submarines. The 688-class of boats was the boat of the time, and they were all broken. One of the enduring traits of all submariners is the realization that "if it is broken, fix it." The summer of 1979 saw me as a very busy manager, with much guidance and direction to "fix it" from my bosses. Into this pressure cooker walked Harry Yockey, Project Manager of the submarine-launched Tomahawk cruise missile. He worked in a Joint Project Office with Air Force and surface Navy guys and gals. Harry asked me what I was doing to make the 688-class submarines capable of firing his cruise missile. My answer was easy: "Nothing."

I had enough on my plate, and I didn't think any newer "stuff" was needed. However, Harry was a classmate, and I must admit the idea of turning a nuclear attack submarine into the wind to launch a cruise missile was a captivating concept. I did not know Harry at the Academy, nor had we worked together before. But, as I am sure many of us have experienced over the years, there is a union that comes into play here. Classmates are special!

There are certainly exceptions, but I do believe there is a bond between all of us that makes us want to help one another. Besides the fact Harry was a classmate, I had spent most, if not all, of my submarine career with a "discrepancy list" in my hands, and here was an opportunity to do something that might be important to the Submarine Force, the Navy, and the country. One might think that a Tomahawk capability would have been a high priority and an area where intense oversight would be the order of the day. Not so—the Washington bureaucracy was not interested. I am not sure why. The cruise missile task was Harry's and mine.

At the start, Harry had a missile designed to be fired from a standard twenty-one-inch submarine torpedo tube. My initial task was to make the 688-class submarines Tomahawk-capable. Once I had adequate systems to satisfy the difficult command and control requirements, I wanted to increase the number of missiles carried on board. Vertical launch from tubes installed outside the pressure hull was the idea. This notion was new, different, and transformational.

One of my thornier problems was convincing our bosses we could add twelve missiles without affecting the noise level or speed of the boat—important considerations on a submarine. To take my back-of-the-envelope ideas and do a first-class concept paper that would address all the issues and have all of the 't's" crossed and "i's" dotted, we needed money. Harry and I went to the Submarine Desk in the Pentagon. The response was to the effect, "What? Are you guys nuts?" and we left with our tails between our legs. Retreating to go to the parking lot, we were marching along the E-wing in a Pentagon corridor when we noticed the Deputy CNO for Undersea Warfare coming toward us. He turned left and went in a door. When Harry and I reached that door, we noticed it was marked, "MEN." We looked at each other and went into the men's room.

That afternoon, I received a telephone call from the Submarine Desk. I was informed funds were located for Engineering work!

Twenty-five years later, the Submarine Force is marketing a submarine that will carry 154 of Harry's missiles. I wonder if there is any notion that Tomahawks launched from vertical tubes were born in a head (MEN'S ROOM) on the fourth deck of the E-wing of the Pentagon, by one Vice Admiral and two Captains from the Class of 1957 while taking care of "business." Teamwork and determination at work!

STARTING A TRADITION
Jim Googe

FOR OUR CLASS, we remember the life-size bronze statue of our mascot, Bill the Goat, in front of McDonough Hall on the Annapolis campus (Yard). We also remember that Bill is sculpted in a fighting pose, as befits the Navy sports mascot, charging forward off his hind legs, horns lowered to ram, with front legs raised, revealing in the anatomically correct fashion, those parts which impressively distinguish a billy goat from a nanny goat. (Understand?)

Well, they tell me now that the statue has been relocated to a different spot in the Yard, and that a tradition has evolved requiring Bill's impressive distinguishing features to be painted by Midshipmen or polished or rubbed for good luck to the same effect before important events—such as the Army-Navy football game or final exams. Frankly, I don't know what they're doing now, but I do know how it started . . . because I started it.

Bill the Goat, the Navy Mascot.

In the winter of my second class (junior year), during my exchange week at West Point, I was admiring the large equestrian statue of George Washington on the parade ground. I noticed the horse was anatomically correct, and prominently so, and was told by a cadet the horse's not-so-private parts were colorfully painted by cadets on important occasions. Among the many dark memories of that cold and dreary place, that was a brighter one I brought back to Annapolis. At the time, I didn't realize how important this knowledge would become.

Fast forward to June Week of that same academic year. Back in Annapolis, a magnificent bronze statue of Bill the Goat, the mascot symbol of Navy athletics, had just been *erected* at the west entrance of McDonough Hall, the home of Navy athletics. The dedication of the statue as a gift to the Academy from the Class of 1915 was to take place on Saturday morning at the beginning of June Week. The ceremony would be attended by many dignitaries, including the President of the Class of 1915 and the Superintendent of the Academy, Admiral Smedburg. All Captains of varsity sports teams chosen for the next year would also be there. As Captain of the varsity sailing team, I would attend.

The statue of Bill, like the West Point horse, was also anatomically correct, and his posture lunging forward and upward, made it abundantly clear that he was a real RAM, ready to smash into his foe.

The day before the dedication, the statue had been carefully covered with a tarpaulin. It was then that a light bulb went off in my head. How wonderful it would be if all those dignitaries at this important ceremony were surprised at the unveiling, by a goat having one testicle painted blue and one painted gold.

The plot quickly hatched, key members of the varsity sailing team were recruited to rendezvous in the shadows near the statue after midnight in dark clothes and watch caps, with flashlights, brushes, and durable enamel paints in blue and gold. We hid as security drove by on their nightly patrols, then dashed out, and quickly under the tarpaulin, the deed was done. Safely back in our rooms, what a joy! We were confident that I, their Captain, would soon be an official witness to the most memorable class gift dedication ceremony ever.

Early the next morning, what a disappointment! Workers had come to set up the venue. When they removed the tarpaulin to install a fancy blue and gold striped unveiling shroud, our handiwork was uncovered. For the next hour or so they worked diligently with steam lances and wire brushes to remove all traces of the Navy's proud colors from Bill's nether regions.

But the time had come. The dignitaries had all said their piece. I was glumly sitting and waiting for the aide to pull the drop cord to reveal an un-resplendent Bill, but I had failed to anticipate what all that steam and wire-brushing would do to burnish Bill's gonads so brightly against the contrasting dark green oxidation on his bronze body. My utter delight was complete when I saw Admiral Smedburg, point at Bill's underside and exchange whispers and chuckles with the Class of 1915's President. There was a great round of applause, and our handiwork was proudly appreciated and properly recognized.

I have since learned that the Plebes, from time to time onward, have been required to polish Ole Bill where it counts before important games and events. It wasn't exactly the tradition I envisioned, but I'll take the credit (along with my teammates, of course) for starting it.

All the above is true because I was there. The following is hearsay, but I like to think it is true. My oldest granddaughter is married to an Army officer and VMI graduate. While at the Army Staff College at Fort Leavenworth years later, she and her husband were exchanging Academy tales at the club when she told my story without attribution. She asked a Naval Academy graduate in the group if he knew who started the tradition. He said, "I don't know; I think it was some guy named Googe." She replied, "He's, my grandfather." So, I'm famous!

PS: It must bring good luck to what we did, because the sailing team (my team) won the National Championship for the first time three weeks later. And the year after that as well!

TWENTY-SEVEN MEMORABLE YEARS IN THE NFL
Bob McElwee

MY TWENTY-SEVEN YEARS in the wide "wild" world of NFL football was an avocation full of memorable and sometimes comical events. Here goes an attempt to share some of them with you and my classmates. (You were my greatest and most supportive fans!)

I was in Berlin, Germany, to work a game between the Chicago Bears, with coach Mike Ditka and Walter Payton, and the 49ers with Joe Montana. The Berlin wall had just come down, and we played in the huge one hundred thousand-seat Olympic Stadium which was jammed to capacity. Here was where Jesse Owens ran his famous one-hundred-yard dash in the Olympics and gave Hitler a taste of what was going to happen to him when he challenged the USA. You could almost see Hitler and Mussolini reviewing troops there and feel the history emanating from the stadium walls. I truly felt my heart pounding in a sense of awareness of the magnitude of where I was and what it all represented in the ominous Nazi regime.

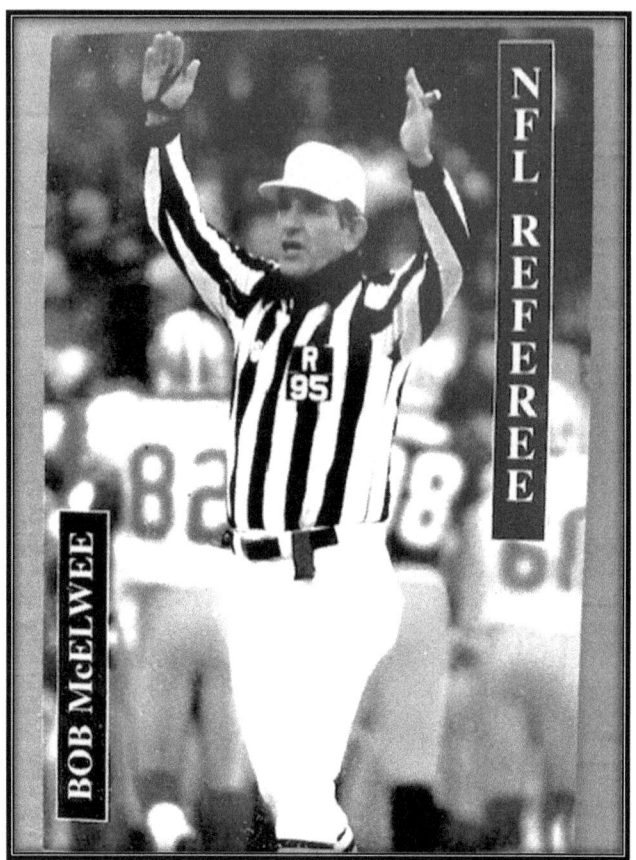

Bob McElwee signaling touchdown.

Just before the coin toss ceremony, they brought to me the beautiful Katerina Witt, the East German Olympic figure skating champion. I was introduced to her, and they informed me she would toss the coin before the game. I replied that would be fine, and just before we went out for the toss, she put her hand on my arm and said "Bob, there's something I think you should know."

I said, "What is that?"

She said, "I am East German, and they don't like me over here in West Berlin. When you take me out there, they are going to whistle. In Europe, that means they are booing."

I laughed and replied to her, "Young lady, don't worry about that for one second, because I have walked out to the center of just about every major stadium in our country, and they booed me all the way out and all the way back!"

I was privileged to have a West Pointer, Al Conway, on my officiating crew for eight years. Our Army vs. Navy bantering aside, we became the closest of friends. Al was a Colonel Blaik Army football player, and he was the Philadelphia Eagles' number one draft choice in 1954. He was a Vince Lombardi kind of guy; tough, strong, structured, loyal, and smart as hell. He would tackle a rattlesnake if it was in his way. If there was trouble, we always knew the other guy would be there for us... no questions asked!

Al and I were working a game in New England years ago. The Patriots were playing Minnesota and on about the Vikings ten-yard line going in, Steve Grogan, the New England quarterback, rolled out to his right and threw a pass into the end zone. I did not see the ball, as my job was to keep my eyes on the quarterback as he was tackled to the ground after releasing the ball. The crowd told me something had happened, and I looked up to see a Viking defensive back, who had intercepted the ball in his end zone, coming back up the field toward me. I sprinted alongside him the length of the field and signaled a Minnesota touchdown. As I turned back and looked up the field, I saw a lot of players still up in the other end zone, and Al running toward me with a sheepish grin on his face. "Guess what," he said. "You're not the only one who has a touchdown on this play. The head linesman has a touchdown for the other team up in the other end zone!"

"A touchdown in both ends?" I said. "That's interesting. What do you suggest we do now?"

Al calmly replied, "Let's stay here!"

"That sounds like a typical hard-headed West Point answer," I said. "Now come with me while I find out what happened in the other end zone."

Our discussion revealed that the head linesman had raised his arms for a touchdown while both the Viking and Patriot players had the ball jointly possessed in the air. The Viking defender had ripped the ball out of the other player's grasp before they returned to the ground and took off for the other end zone one hundred yards away with an interception. And this was before the days of instant replay! I decided to award the Patriots a TD based on the head linesman's "first" TD signal, which ended the play-by rule before the interception occurred. So, there you have it ... two touchdowns on the same play, an NFL first! Needless to say, I had received very little help from my West Point friend, and we slipped hurriedly and quietly out of town!

Probably my favorite player of all time was Walter Payton. Not only was he a great player, but he loved to play the game. Walter also believed in having fun on the field, and because of him I always had to be ready for anything when I worked the Bears. Officiating involves many disciplines. After each play and before the next play, the referee must run through a series of checks. Game clock, play clock, downs, distance, substitutes, eligible receivers, etc., etc. When I was working with the Bears, I always had to add one to the list ... checking my back pocket to make sure my penalty flag was still there. Walter would bump up against me while I was retrieving the ball from him after the play and my flag would disappear from my pocket. When I checked and found it gone, I would mosey on up to the back of the Bear's closed huddle, stand right behind #34, and tell Walter I wanted my flag back. He would take the flag out of his pants where he had stuffed it and, without disturbing anyone in the huddle, hand it back to me while the quarterback was calling signals. With my flag in my pocket, I was ready for the next play.

I was working a Bears game on a rainy, muddy day in Soldier's Field. My umpire was a grizzled old gladiator from the University of Georgia and the Steelers' Frank Sinkovich. As the game progressed, Frank came to me three or four times and told me his shoelaces kept coming untied and he couldn't figure out why this would happen just because the field was muddy. We never figured it out until the next Saturday when we were reviewing the previous week's game film in the hotel room as a part of our weekly preparation for Sunday games. They were the old sixteen-millimeter films which were hard to see in detail on a clear day, much less a rainy, muddy day. But I thought I saw something after Walter was tackled on a play and told the guys to stop the film and rerun it in slow motion. And sure enough, there was the answer. As Frank Sinkovich came over and reached for the ball from Walter as he lay tackled on the ground, out came this little hand, grabbed Frank's shoelace, and jerked it loose. And this was going on all day!

One of the lessons I took from Walter Payton I try to convey to young people today, is to find a vocation or a life's work you enjoy. Whatever the job, if you love your work, you will probably be successful. Tragically, we lost Walter too young to a rare disease, but my memories of those Sunday afternoons in Soldiers Field will stay with me forever.

Early in my career, I was assigned to a playoff game at Shea Stadium in New York the day after Christmas, between Buffalo and the New York Jets. I was sitting in the locker room, dressed and ready to go on the field, when three large men dressed in dark suits, ties, and topcoats came toward me and introduced themselves. They were FBI agents and they informed me there had been a death threat on Richard Todd, the Jets's quarterback. "We're not telling any of your other officials," they said, "but we wanted you to be aware of it." *Nice Christmas present*, I thought, but I had work to do and took the field for our pregame routine. I was no sooner on the field than Richard Todd came running toward me. Looking for another direction to turn, I realized I was trapped and met him at the twenty-yard line.

"What are you going to do when they start shooting?" he asked.

I answered "I don't have all that padding and body protection all you quarterbacks wear, just this striped shirt which won't stop anything. So, I'm going to stand right behind you!"

As you probably know, in the NFL the referee is responsible for the quarterback at all times. Wherever he goes, we'd better be right there with him to protect him. But not this day after Christmas! If Todd ran past the line of scrimmage, I watched him carefully, but from my position far behind the line. If the Jets called a time-out, I found a way to discuss something with the official who was farthest away from Todd. If Todd wanted to talk to me, I was sure to keep him at shooting range! History tells you nothing happened, but it certainly was one of my more interesting days!

People often think the referee has a relatively easy job. Just show up, make the calls, and get out of town in a hurry. Well, here's an example that will make you wonder why a guy would be crazy enough to do what I did for twenty-seven years.

I was assigned to a playoff game between the LA Raiders and the Denver Broncos in the Coliseum in Los Angeles. John Elway and Jeff Hostetler, (the Raiders quarterback), were lighting it up, and at halftime, it was 20–20 . . . a real barn burner.

Early in the third quarter, Hostetler rolled out toward the Raiders' bench (with me on his tail) and was blasted by a Denver DB right at the sideline. The impact knocked Hostetler right into the Raiders' bench, and out they all came after the Denver players. The contact had been legal, right at the sideline, but the Raiders players didn't want to hear it. There had been bad blood between them in a game a couple of weeks ago, and a full-blown melee ensued.

After we got things settled down, the back judge told me that a guy in civilian clothes came off the Raiders' bench and was in the middle of the field in a fight with the players. Realizing that TV time was probably worth about a million bucks a minute and that the game was being held up, I wanted to get out of this mess and resume play.

"I'm telling you, Bob, I think he's a Raiders player in civilian clothes," the back judge said.

"I can't tell if he is or if he's the hot dog vendor," I said, "and don't have time to find out. So, get back to your position on the double because I'm going to blow the whistle and we're going to play football."

After the game, we found out he was a Raiders player and during the game, the LA police arrested him, took him downtown in the paddy wagon, and locked him up for assault and battery.

Later in the game—the same game, believe it or not—I am standing in the end zone waiting for the TV to return. There are over one hundred thousand people in the Coliseum, but we are in a TV time out and the crowd is relatively quiet. Suddenly, I hear this buzzing, like an electric motor. I look up, and here is this guy in an ultra-lite kite with a motor on his back, circling like a hawk high above the top of the stadium.

I immediately go to Art Shell, the Raiders coach, and Wade Phillips, the Denver coach, and tell them to get their players off the field. Then I go to security and tell them to call the LA police. In the meantime, this guy is hovering around over the stadium, and I'm wondering if he's going to launch a hand grenade or what he's going to do. Finally, after about five minutes, a police chopper appears over the top of the stadium, chases him until he disappears behind the stadium wall, and we resume play. We were told later that this was "Kite Man," the same guy who dropped into the ring before a Sonny Liston heavyweight championship boxing match in Las Vegas. He also had himself dropped into the top of the Parliament building in London, naked as a jaybird, and the Queen banished him from England! (ESPN has done a story on his life, which you may have seen.)

So, there you have it—two different guys, both involved in some way in my football game, both thrown in the paddy wagon and taken downtown and locked up, and we're still playing football! How's that for a Sunday afternoon's work?

SNEEZING CRICKETS
Bill Baab, Friend of USNA Class of 1957

Y EARS AGO, I was fishing in Merry Brothers' Brickyard Pond when a violent thunderstorm sent me paddling back to shore, and I huddled beneath my overturned aluminum Jon boat to wait for the storm to pass.

While waiting out the storm, I got an idea, and when I returned home, developed it into a harrowing story for my Outdoor Column in *the Augusta Chronicle* newspaper (Georgia).

I was fishing for bedding bream and the bite was very slow, not catching a darn thing. After coming ashore to take shelter from the storm, I placed the cricket boxes in the ice chest so the bugs wouldn't get wet.

After the storm passed, I shoved the boat back into the water, paddled back to my secret spot, took my cricket cage out of the ice chest, and started fishing again. I was fishing a cricket with no sinker beneath a cork bobber.

I heard a tiny sneeze, saw the cork disappear, and soon hauled in a fat bream. What could have happened? The crickets had caught colds in the ice chest, their little sneezes woke up the bream, and soon I was catching fish like crazy — had my limit of nice fish in no time.

This was such a nice story, I wrote a fishing column including my imaginative sneezing crickets, and it amused my *Augusta Chronicle* readers (back in the 1960s). Are you laughing?

SNEAK ATTACK ON A JUNK
Charlie Hall

———

ONE LATE AFTERNOON, I had the Fire Team on our Huey helicopter out looking for some action and was talking to the US Army Liaison to see if they had potential targets. All was quiet and serene in the operations area—there was no traffic on the rivers and canals because of a curfew and nobody was allowed on the water after the curfew. In our Huey Helicopter, we flew down the coast, and while waiting for some intelligence to employ as a target, we came across a medium-sized junk that was big enough to have some sort of superstructure that could carry a fair amount of cargo.

We spotted him from a distance because we were flying at about two thousand feet, so we were essentially immune to random small-arms fire from the ground. Consequently, we could see quite a way around us. Since I was talking to the US Liaison, it was easy to determine the craft had no business being where it was and, besides, curfew was fast approaching, and the craft would be in violation almost immediately. The US Liaison cleared us to attack the craft and so I chose a sneak attack aimed at not letting the people on the junk know they were in trouble until it was too late to avoid the attack. I told my trail ship Commander to maintain altitude and planned to drop down and fly at a low level on the river until I made the last turn to roll in on the junk. Trail ship was to keep me informed about the junk's movements and join in after I made the first run at the target.

I armed my rockets and briefed the gunners I would fire rockets first as we approached but I expected our starboard side fifty-caliber machine gun would be the best weapon to neutralize the target. My trail helicopter was able to keep the target in sight without altering his course and so, the people on the junk were likely not aware of my low-on-the-water approach, and were continuing their way when I rounded the last curve and fired my rockets. The rockets went somewhere short and long and the real attack came when I turned to port and climbed to get some altitude so my Leading Petty Officer Zirfluh could work out with the fifty, he was a very accurate gunner. He could hold the gun on target over a small sampan and just cut that boat in half from an altitude of two thousand feet and that junk was a big target. He poured enough ammunition into the junk that it immediately caught fire, and it burned so bright I was sure we had intercepted a cargo of C-4 explosives or something similar.

Now, let me tell you about C-4 explosives. You find it in various applications, one of which is claymore anti-personnel mines. Those were used for defenses of outposts and other such friendly locations. Claymores are very safe because they must be set off by a detonator. Claymores were often disassembled so US ground forces could use the C-4 to heat their rations. It burns hot but will not explode from just lighting a match to a small pinch of it. But a small pinch is enough to heat your supper. So, when that junk caught fire with that very bright flame, I was pretty sure the cargo was C-4 or equivalent, bound for the Viet Cong or other enemy units somewhere in the Delta.

As soon as I started firing, the junk turned and headed for the nearby beach. It ran aground, and the crew jumped and ran for the Nipa Palm swamp. About this time, the trail ship joined the attack and was also pouring gun and rocket fire into the craft.

We remained on target until we were low on fuel, at which point we headed for the nearest fuel/ammo site. We rearmed and refueled and since we had no other targets, we returned to the ship. On the way home, we could see the burning junk from several miles away. It burned for quite a while and I have no reason to believe it was not completely destroyed. And I continue to believe that a huge amount of explosives headed toward attacks on friendly forces went up in smoke that evening. One for the good guys!

SECRET SHENANIGANS
Fritz Steiner

BILL BRYANT, FRED BRADLEY, AND I were the Class of 1957s contingent in the submarine *Razorback* during 1961–62. I hope Bill will forgive me for telling this one about him.

One day, while in port in San Diego, I was the Duty Officer. Bill, who was then the Registered Publications Custodian (secret stuff), had made a trip to the base Registered Publications Security Office to draw our latest issues. I saw him come back with his "draw" but didn't pay much attention. I was the Engineer and had enough problems of my own. For some reason or other, Bill had to leave the boat early that afternoon, and in somewhat of a hurry—but that was none of my business.

Some hours later, I went into the six-man stateroom to wash up for dinner and to ready myself for the impending battery charge. I couldn't help but notice the curtain on Bill's bunk wasn't completely pulled shut. There on his bunk were several unopened, but easily recognizable classified documents. I guessed that in his haste to leave, Bill had forgotten to put them away in the Registered Publications safe. I picked them up and stowed them in the Duty Officer's safe in the Wardroom.

Early the next morning, Bill returned to the boat and earnestly began looking all over the Forward Battery Compartment for something, but he was saying nothing. If it had been, say, his wallet he'd misplaced, I'm sure he'd have asked me and the stewards if we'd seen it. But of course, THAT wasn't what he was looking for, and he sure as Hell, didn't want to let anyone know what it was.

I'd intended to tell Bill as soon as he came aboard, but he was already searching before I realized he was there. Since I knew what it was he was looking for, I decided instead to watch him for a bit. After a few minutes of fruitless search, Bill began to look panicky. (None of us had ever seen Portsmouth—a military prison—but at times like this, we could imagine what it looked like—*bad*.) He had suffered enough. I went to the Duty Officer's safe, opened it, took out the missing secret material, and said, "Bill, is *this* what you've been looking for?"

Bill almost passed out with relief. Until this writing, only he and I have ever known about this. Now the world knows. OOPS!

Hong Kong Security

During the fall of 1957, my first ship USS *Lenawee* (APA-195) was assigned duty as the Station Ship in Hong Kong. Among other things, the station ship picked up the communications guard responsibilities for visiting Seventh Fleet ships and other naval entities in port for rest and relaxation. Thus, we had the communications duty for all ships and visiting VIPs.

One morning, I'd just been relieved of my duties as Duty Officer for the early morning quarterdeck watch and was on my way to breakfast and a nice nap afterward when the Communications Officer intercepted me. It seemed the Chief of Naval Operations (top Admiral in the Navy) and his entourage were in Hong Kong "inspecting the defenses." The Communications Officer said that, after breakfast, I was to go to the Gloucester Hotel on the island, go to a certain room on the top floor, and pick up something the CNO wanted to be returned to the ship for safekeeping.

Naturally, I replied, "Aye, aye, sir."

So, I went with a driver from our Shore Patrol Security group to the hotel, caught the elevator up to the top floor, and found the room. The door was ajar, so I knocked and entered. An officer considerably senior to me (Hell, everybody was senior to our Class of 1957 back then) was busy tying his tie and seemed to be running late. A Chinese room attendant was making the bed.

I said, "Sir, I'm Ensign Steiner from the Station Ship reporting as ordered to pick up something from the CNO for return to the ship."

"*Harrumph*, yeah, uh, Mr. Steiner, come in. It's over there on that bedside table," He grumbled.

"Aye, aye, sir." I went to the table, passing the room attendant near it.

Ah, yes. There "it" was: four messages (two marked TOP SECRET, one marked SECRET, and one marked CONFIDENTIAL) all in plain English and in plain sight. As I picked them up, I gulped. I'd never seen anything "TOP SECRET" before in my life. I folded the messages and put them inside my shirt.

The officer said, "Steinmetz, take those out to the Station Ship as soon as you can."

"Uh, sir, it's Steiner," I corrected.

"Yeah, Steiner, right. Okay, get going."

"Aye, aye, sir." I exited and rode the elevator back down to the ground floor with several Chinese in it. As I stepped out in front of the Gloucester Hotel to await my driver's return, I was apprehensive, to say the least. Here I was, standing on a crowded Hong Kong street (but I repeat myself) in midmorning with unencrypted hard copies of highly classified messages tucked in my shirt.

I looked to my left—oh, no, the offices adjacent to the Gloucester Hotel were occupied by the Chinese Communist airline.

At last, my driver arrived, and I got the messages out to the ship and locked in the safe.

Thinking back, that had to have been a huge security gaffe by that officer. He never asked me to prove I was who I said I was. He had plain text "Top Secret" and "Secret" messages lying loose on a bedside table in plain sight of a foreign national whose loyalties were completely unknown. The officer wasn't paying attention to anything but tying his tie.

I suppose I should have asked him who he was, too, but I was an intimidated Ensign and didn't know any better.

Let's just say we got away with it. Whew!

SHARKS IN THE PACIFIC
Wilson Whitmire

IN JUNE 1962, I was a weapons officer aboard the diesel submarine *Bugara* (SS-331) which was three days west of Midway Island and headed for Yokosuka, Japan. The seas were glassy calm. An object was spotted broad off the starboard bow at a range of about six miles and upon investigation, it proved to be a steel sphere about eight feet in diameter that was probably one of many used to buoy a cable net strung across a Japanese harbor entrance during World War II.

As is common in the open ocean, anything drifting on the surface attracts a variety of fish, and this sphere was no exception, hosting small fish near the surface and dolphinfish (mahi mahi) farther down. The sub's Skipper, Commander Larry Marsolais, and I were both avid fishermen but had no luck catching the dolphin.

I grew up skin diving and always carried my gear with me, so the Commanding Officer permitted me to attempt spearing the fish. After about thirty minutes, I had two nice twelve- to fifteen-pound fish tended on a line tied around my waist and was going after a third, when three large sharks came up from the deep and began circling me, dorsal fins protruding above the surface.

I immediately cut the fish loose, and they were inhaled by two of the sharks. I had hoped the third didn't feel slighted! At this point, the submarine was about two hundred yards away, and I began a slow swim toward the boat. When a shark would get too close, I would attempt to push off using the butt end of the spear gun.

Occasionally, a shark would scrape me while swimming past and that's when I first learned how rough and sandpaper-like shark skin is. Between my swimming and the submarine's maneuvering, I was alongside after about ten minutes and then was hauled up by line over the tank tops. Just as I made it to the main deck, a fourth shark, undetected to this point, came up almost vertically from the deep, broke the surface, and rolled halfway up the tank top I had just vacated.

This shark was estimated to be twelve to fifteen feet! Afterward, Larry Marsolais said his first thought when he saw me encircled by shark fins was how he was going to have to explain to higher authority the loss of a junior officer to sharks in the middle of the Pacific Ocean! That encounter ended my spearfishing on the submarine.

In later years, Larry and I had many wonderful fishing experiences together, both in the States and in the Sea of Cortez in Baja California. On more than one occasion after hoisting a few, we would reminisce about the shark encounter and kid about who was more worried, me for my life or Larry for his Naval career.

What became of the sphere? Attempts to puncture it with a fifty-caliber machine gun were futile, so we sent out a hazard to navigation message. The sphere may still be drifting around, attracting offspring of those fish from 1962.

SHOT DOWN OVER ENEMY TERRITORY
Larry Bustle

ON FEBRUARY 15, 1968, I reported to the 480th Tactical Fighter Squadron, 366th Tactical Fighter Wing, at Danang Air Base, South Vietnam. I had just recently upgraded to the F-4 Phantom II aircraft, and I was assigned to combat duty as an F-4 pilot. My reporting date, February 15, 1968, was just two weeks after the North Vietnamese Tet Offensive began, so everyone at Danang was still a bit jumpy. North Vietnamese and Viet Cong soldiers had penetrated the base and were repelled after a serious gunfight on the base proper.

I was one of the very few pilots in the wing who had been a fighter pilot throughout his career to that point. Most of the pilots in the wing were "retreads" from airlift, air training command, or bomber career paths and many of them had reached relatively high rank, some by Strategic Air Command's spot promotion program. I was also a graduate of the USAF Fighter Weapons School at Nellis AFB, Nevada, so I was rapidly pressed into duty as a flight lead and eventually assigned to work in the Weapons and

Tactics section of the wing. I was the fifth ranking major in my flight, which shows how high the rank structure was in the wing. There were four flights in a squadron at that time, with a total of about ten to twelve people in a flight.

I flew all types of missions, including close air support, air interdiction, and air-to-air, in South Vietnam and the lower part of North Vietnam (Route Package 1), although my wing's mission at that time did not include the upper reaches of North Vietnam where the Migs were located. By September of 1968, I had 130 total combat missions, including sixty-eight over North Vietnam. In many of our missions over North Vietnam, we worked with the Misty Fast FACs (Forward Air Controllers). The Mistys flew two-seat F-100F Super Sabres, usually with two pilots in the aircraft to increase the number of eyeballs looking around. Their mission was to get to know the "Pack 1" terrain intimately and search out and find trucks ("movers"), barges, surface-to-air missiles (SAMs), anti-aircraft artillery (AAA), and any other targets that might be considered of high value. Then they would work with strike aircraft and designate the targets using smoke rockets. As they searched for targets along the Ho Chi Minh Trail and in the southern part of North Vietnam, Mistys would usually try never to get below 450 knots or below 4,500 feet above the ground, and they tried to keep changing altitude and direction continuously so as not to provide an easy target for Vietnamese AAA. If the Mistys decided to go below 4,500 feet to get a better look at a possible target, they would up their speed to five hundred knots or above.

After entering North Vietnam on a strike mission, if for some reason we could not attack our previously assigned target, we would contact Misty and see if they had a target for us. We would be flying in a flight of two F-4s usually. The Misty would direct us to the location of a target, describe it to us along with a recommended direction of attack, brief us on any defenses he had spotted in the area, and then we would set up to attack it if we had the proper weapons. We would sometimes wind up in North Vietnam with napalms, high-drag retard weapons, and a twenty-millimeter gun pod, all of which are generally regarded as low-altitude weapons, so we would probably not attack a target in North Vietnam in a high-threat area (i.e., one reported to have a lot of AAA). If we had slick bombs (those without retarding fins or chutes) we would attack the target Misty had found for us. As strike aircraft, we also tried to fly faster than 450 knots and above 4,500 feet and to keep the aircraft moving around as much as possible. The policy for missions over North Vietnam was to dive bomb at a minimum of a forty-five-degree dive angle, so a dive bomb pass would usually start about ten thousand feet above the ground. When we rolled in on a target, we would try to stabilize the bombsight on the aim point, get the dive angle and airspeed stabilized on the desired numbers, then pickle and pull out so as not to go below 4,500 feet. With the accuracy of the gun sight of that day, we could not dive bomb consistently with great accuracy, although our accuracy improved as the number of missions increased. This policy also posed a problem for those times we wound up in North Vietnam with low-altitude weapons, because napalms, retard bombs, and twenty-millimeter cannons were not very effective when delivered at a forty-five-degree dive angle.

The Misty FAC mission was one of relatively high risk, and they lost about thirty-four aircraft in the two-plus years of their program, generally to AAA gunners. Since the F-4 was a two-seat aircraft, it was decided to set up a project to test the use of the F-4 as a fast FAC. The number of two-seat F-100s was decreasing and we had lots of F-4s at that time. We were flying with weapons systems officers (WSOs) in the back seat of the F-4. These aircrews were either another pilot or a navigator, both of whom would have specialized training in the radar and other systems that had the controls in the back seat. They were also called GIBs (guys in the back). Four aircraft commanders (front seat pilots) and four GIBs were selected to set up this F-4 fast FAC program which was to be called the Stormy Fast FAC program. Steve Ritchie, who later became the Air Force's only pilot ace in Vietnam (five Mig-21 enemy aircraft kills), Doug Patterson, one other pilot whose name escapes me, four GIBs, and myself were selected to be in the initial cadre.

The four of us were sent to Phu Cat AB, the home of the Mistys, to fly with the Misty pilots and learn the techniques of the job. Later, the Misty pilots who flew with us would come to Danang to fly three missions in the back seat of the F-4, presumably to advise on its suitability for the fast FAC mission. I flew five missions with Wells Jackson, a guy from the Albuquerque area, and he taught me many of the fast FAC techniques, including how to see targets from 4,500 feet. It was a big surprise for me to find that you could see a lot of small details from that altitude. He showed me how to recognize a thirty-seven-millimeter AA gun site and a fifty-seven-millimeter gun site, and I believe we might have seen an eighty-five-millimeter or one-hundred-millimeter gun site, but I'm not sure. The sites were laid out in a standard configuration and when the gun site was inactive, the gun crews put branches over them. When active, the branches were removed, and the site stood out very distinctly.

On one mission, he showed me how to tell what caliber the guns were by rolling in on a gun site from about twenty thousand feet. Immediately after we rolled in, we could see flashes from the guns in the site, and by counting the number of rounds a gun fired, you could determine its caliber, the reason being that different calibers of guns had different sizes of clips for their rounds. As I recall, a thirty-seven-millimeter gun had six rounds in a clip, and there was a distinct pause between clips while the gunner reloaded. In this demonstration, as soon as we rolled in, the guns started shooting, so we immediately pulled up and discontinued the maneuver. There was a well-known fighter pilot rule that you never get in a pissing contest with a gun; the aircraft usually loses.

On another mission, he showed me why we should never get too close to the ground at a slow airspeed. All along Route 1, which paralleled the coast, there were small white spots where the sand had been disturbed. I asked Wells about it, and he said they were spider holes and people were in each one of them. He pushed the F-100 up to five hundred knots, and we dashed across the highway at approximately one thousand feet. As we crossed the highway, he told me to look back and I saw muzzle flashes coming from each of the spider holes. We were well above their maximum effective range and were so fast they would have had to be very lucky to hit us, but it was still very sobering.

At the time I was flying the five missions with Wells Jackson, I had about 1,800 hours of flying time in the F-100, probably a lot more than Wells had. On one flight, I convinced him to let me fly in the front seat of the aircraft, although our five flights were supposed to be in the back seat. Our mission with me in the front seat went very well, and I had a great time back in the "hunt" after several years. You use a lot of fuel quickly in the high-power settings required for a fast FAC mission, so we finally had to go out and hit a tanker to top off our tanks before we could continue the mission. The F-100 used a probe and drogue air refueling system wherein the pilot of the F-100 flies behind the tanker aircraft and inserts his wing probe into the drogue, which is being trailed behind the tanker. Since I was current in the F-4, which used a boom receptacle system (F-4 flies in position behind and below the tanker's boom; the boomer aircrewman flies the boom and inserts it into the F-4's receptacle which is on the top of the fuselage behind the rear cockpit), and it had been some time since I had air refueled in the F-100, I didn't hit the drogue the first time. Meanwhile, our fuel remaining continued to decrease, not to an emergency level, but to one of concern. So, Wells decided he'd better take over and get us some gas and stop fooling around. He quickly plugged in, took on a full load of JP-4, and we headed back for Pack 1.

With five missions under our belts, we went back to Danang and proceeded to start our Stormy program. Steve Ritchie got the honor of flying the first Stormy mission, and I flew my first Stormy mission a day or so after Steve. We would be matched up with a GIB (guy in back), depending on who was available, since the GIBs in the Stormy program were hand-selected also. Although assigned to fly the Stormy missions, we continued to fly our regular strike missions. On September 11, 1968, I was scheduled to fly a Stormy mission and was paired up with Lt. Rick Van Dyke, a young Air Force Academy graduate. Rick was a sharp pilot and was the victim of the Air Force policy at that time to assign pilots just out of pilot training to the back seat of the F-4. Toward the end of my time at Danang, they started putting more and more navigators in the back seat, but on September 11, 1968, my GIB was a pilot.

We briefed the mission for a midmorning takeoff, got the weather and an intelligence briefing, and proceeded to launch for a routine Stormy mission—or so we thought. We were over Route Pack 1 by eleven and started looking for anything that looked like a target. Finally, around noon, we spied what appeared to be a truck. I called in-the-blind to see if there were any strike flights around and a flight of four F-4s from Thailand answered. I briefed them on the location of the target and advised them to proceed to the general target area and hold while we went out to find a tanker and refuel. We quickly found a KC-135 on Tan Anchor off Route Package 1 and took on a full load of fuel. We hurried back to the target at about ten to twelve thousand feet, using our inertial navigation system to relocate the target. We contacted the strike flight and briefed them on the target.

I don't remember briefing them on any enemy defenses in the area, but it turned out there were plenty. I advised the strike flight I would mark the target with a white phosphorus (Willie Pete) rocket, and Rick and I, as Stormy 01, rolled in to fire a marking round at the apparent truck. I fired the marking rocket at about 8,500 feet above the ground and immediately started a pull-out. During the pull-out, we experienced a very loud explosion and my cockpit filled with smoke. I continued the pull-out and started talking to Rick, advising him to "hang in there, because we're going to make it out over the Gulf." I pushed the "panic button" and cleaned off all the external stores, including two fuel tanks, the pylons and launchers for the marking rockets, and the twenty-millimeter gun pod, which was attached to the belly of the aircraft. I don't believe we ever got below about 6,500 feet in the pull-out, and I turned the aircraft to the west, went into the afterburner, and started climbing. I remember thinking I had missed seeing the gun site that was obviously down there, saw no puffs of smoke from flak, but a gun site *was* down there, and he was a good gunner to get me at our high altitude. This target had turned into a so-called flak trap, and we were the victims.

Shortly after the pull-out was complete, we experienced another equally loud explosion. All the while I was talking over the intercom to Rick and encouraging him to hang in there with me. In the F-4, there was an ejection sequencing system that allowed the GIB to select one of two options: the normal option was the Command option, which ejected the GIB first and the pilot in front second at a preset time sequence if the pilot in front initiated the ejection. This was done to protect the GIB from the blast of the seat's rocket motor if the pilot were ejected first. If the GIB initiated the ejection in the Command option, he would be ejected and the pilot in front would not be ejected, unless he subsequently initiated it. The second option, Command Override, would eject the GIB first and the pilot in front second, **if the GIB initiated the ejection!** As I recall the events from thirty-six years ago, I wasn't sure which option Rick had set up, so I wasn't taking any chances and was talking to him to keep him in the airplane. I didn't want us to eject over North Vietnam under any circumstances. As it turned out, I was talking to myself the whole time because the second loud explosion I'd heard was the sound of Rick's ejection.

The smoke was so thick in the cockpit that I could not see the instruments on the panel in front of me. I could, however, see out the side of the canopy so I was anxiously looking for the coastline so I could get out at least five miles so the bad guys couldn't get to me very quickly. I couldn't breathe using the normal oxygen system, so I switched to 100 percent oxygen, then to emergency oxygen, and in those positions, the flow was cut off entirely. So, I started holding my breath as much as I could. Meanwhile, we were climbing and accelerating in the afterburner. I tried to turn around in my seat enough to see Rick in the back seat, but the smoke was too dense and my harness too tight. Usually, you could just barely see the back seater if you twisted around far enough.

I made a Mayday call, and the strike flight we had intended to put in on the target replied. They said they had us in sight and were following us out to the Gulf of Tonkin. I watched the shoreline slide by and when I thought we were at least five miles offshore, I came out of the afterburner and started thinking about ejecting.

Many times, I've reflected on my decision to eject from the aircraft, and I've concluded that it was the best decision. The aircraft was on fire, I was gasping for air, and I couldn't see the instrument panel. I could have jettisoned the canopy in hopes some of the smoke would have cleared, but I remembered the F-4 had a bad history of sucking smoke and flames up into the front cockpit when the front canopy departed the aircraft while the aircraft was on fire. So, I decided I had to get out of the aircraft. Unfortunately, I didn't wait long enough after coming out of the afterburner so the aircraft could slow down, and I ejected going an estimated five hundred knots. Thus, Rick and I became the first Stormy aircrews to get shot down.

I don't remember anything from the time I pulled on the lower ring with both hands to initiate the ejection sequence until a violent jolt when I found myself hanging in the chute. I knew immediately I had hurt myself badly because both of my shoulders hurt, and both legs were swinging the wrong way (sideways . . .). My helmet had been pulled off by the ejection wind blast. I had plenty of strength, despite my injuries, so I pulled out my survival radio and spoke to the F-4s that were following me. I believe they told me they had me in sight, and I told them about my injuries. Looking back, I well described all my broken bones and injuries. I tried to activate my survival kit which was in the seat pack attached to my chute harness, but it would not deploy so I pushed on the connecter and the seat pack fell away. That meant I would not have a life raft once I landed in the water. I remember asking the strike flight to make sure no bad guys were able to get out to me. I tried to land with my face into the wind which required pulling on the harness to rotate myself. We were trained to try to land in the water with our faces into the wind so that the parachute and all of its risers would fall behind us and not get entangled with us. Trying to pull on the risers hurt badly so I stopped trying that. Next, I was going over in my mind the things I had been trained to do in such a situation, and I almost had a heart attack when I remembered I had to activate my LPU (life preserver-underarm). It's a good thing I remembered because I would have sunk to the bottom of the Gulf otherwise.

I landed in the water and in about five to ten minutes, there was a Navy anti-submarine warfare (ASW) helicopter hovering nearby. While waiting for the rescue helicopter, I noticed a small trickle of blood coming from my left elbow, and I thought to myself that it would be a real shame to survive a shoot-down only to be eaten by one of the notorious great white sharks that are in the Vietnam waters. Fortunately, the sharks didn't find me and the ASW helicopter put a PJ (parachute jumper) rescue guy in the water and tried several rescue devices to pick me up; including the horse collar, a stretcher-type device, and a thing I call the bird cage. Even though the Navy PJ tried valiantly to help me, I couldn't help myself enough to get into any of those devices, and in the meantime, the PJ was having a very tough time getting me free from the parachute riser lines tangled around my legs. He told me later he had worn out the blade of his hook-bill knife cutting the lines.

After about thirty minutes in the water, I noticed there was an Air Force Jolly Green helicopter hovering nearby, watching the situation. I pulled my survival radio up out of the water by its lanyard, drained it a bit, and it miraculously worked. I called on the guard channel for the Jolly hovering nearby: "Jolly Green hovering, this is Stormy 01Alpha (Rick was Stormy 01Bravo), please come land on the water and pull me in the side door. I can't seem to get into the rescue devices, I'm cold and I'm beginning to hurt bad."

In a few minutes, he landed on the water about fifty to seventy-five feet away from me, put his PJ in the water, and the two PJs were able to get me to the side of the Jolly Green. I remember looking up in the cockpit at the two pilots in the Jolly, and as luck would have it, they were two of my roommates from Danang, Don Olsen, the aircraft commander, and Wendy Shuler, the copilot. I mouthed the words, "I love you," and my whole attitude about helicopters and the people who fly them has been different from that time on. The PJs pulled me in the side door of the Jolly, got me into a litter, took my G-suit off, and strapped me in. When they pulled me in the side door was the only time I ever remember crying out in pain. It hurt until someone gave me a shot of morphine. I asked them about my GIB, and they said a search and rescue effort was going on onshore for him. That was the first time I knew he was not with me in the aircraft all the time. They asked me where I wanted to go, to the Navy hospital ship (can't remember whether it was USS *Repose* or USS *Sanctuary*???) that was offshore near Dong Ha at the DMZ, or to the AF hospital at Danang. It probably would have been much shorter to go to the ship, but I opted for Danang.

I was pretty much sedated for the next few hours, but I remember the wing commander, my squadron commander and ops officer, and lots of the guys from my squadron coming to see me. The next day, I remember being in a ward where they told me there was a captured and badly injured Viet Cong soldier down the hall. I was medivaced to Tachikawa AB in Japan, where my injuries were repaired. I spent six weeks at Tachikawa with what they call "butterfly splints" on my two arms and hip-to-toe casts on each leg. When I was narrow enough to fit in a litter on a C-141 medivac aircraft, I was flown back to the US. I eventually wound up at MacDill AFB, Florida, close to my wife and family who were living in Bradenton forty-five miles south of MacDill.

I developed a staph infection in my right shoulder and had to undergo some drastic treatment to get it under control. As a result of the infection, I lost some bone in the shoulder joint and some of the muscle and other tissue nearby. It finally healed up and, approximately six months from the day of the event, I passed a flying physical at MacDill and another one a few weeks later at Brookes AB in preparation for my next assignment as a student at the Air Force Test Pilot School.

While recuperating in the hospital at MacDill, I received an envelope with several pictures of the target area we were working on that day. The photos, which had been recently declassified, had been taken by Dick Rutan, a Misty pilot who was flying in that same area on the day of our incident. The photos clearly showed a thirty-seven-millimeter site in the area I had missed and noted the spot where he saw a parachute on the ground, which was undoubtedly Rick's. Dick Rutan retired from the Air Force and later participated with Jeanna Yeager in the first unrefueled flight around the world in the Voyager aircraft which was designed by his brother, Burt Rutan.

When the POWs came home in 1973, I met a fellow whom I had been told had some first-hand information about what happened to my GIB. Art Hoffson, now retired as a Colonel, told me he had been shot down in the same general area approximately ten miles east of the cities of Dong Hoi and Quang Khi about two weeks before our incident. He was being held prisoner in a cave nearby while the North Vietnamese collected enough prisoners to transport them to Hanoi when Rick Van Dyke was brought into the same cave. Rick was in and out of consciousness and was rational only part of the time. The story he related to Art Hoffson was that his aircraft had been hit, set on fire, and crashed near the target area. He thought his front seat pilot, me, had gone in with the aircraft. Rick told him there was fire all around him, so he ejected from the aircraft right over the target area. Rick had suffered a broken thigh, caused either by the anti-aircraft hit, the ejection, or the parachute landing. It was a compound fracture since it was gangrenous. One day, they took Rick out of the cave and told Art they were going to take him to see a doctor. Rick never came back, and the guards told Art that "he die." Rick's remains were returned in the late seventies, and he is buried in Arlington Cemetery.

I completed a twenty-seven-year career in the Air Force and retired as a colonel in 1984, and am living happily ever after with my wife, Edie, in Palmetto, Florida.

BUILDING A SUBMARINE
Peter Boyne

ONE OF THE HIGHLIGHTS OF MY SUBMARINE career was assignment to the commissioning crew of a ballistic missile submarine. My tour on USS *Sea Owl* (SS-405) was cut short when I received orders in November 1964 to go to Navigation School in Dam Neck, Virginia. Upon completion of the three-month course, I was ordered to the pre-commissioning crew of the USS *James K. Polk* (SSBN-645)(Blue) whose commanding officer was Captain Frank D. McMullen. I was to be the boat's first Navigator.

Polk was being built at Electric Boat in Groton where her keel had been laid in November 1963. The *Jimmy K.*, as she became known, was the thirty-sixth ballistic missile submarine to be built in a fleet of SSBNs known as *41 For Freedom*. She was designed to carry sixteen A-3 Polaris missiles.

SSBNs have two crews—Blue and Gold—which allows the boat to remain at sea and on patrol at least 80 percent of the time. While on the building ways, the Blue and Gold crews worked as a team. Offices and crew workspaces were located on an adjacent barge. Days were long, in addition to shift work and nighttime hours. To support our unique navigation system, a new periscope, type 11B, was installed. It allowed for very accurate star acquisition. I spent many nights "shooting stars," except when the skies were overcast. Many a night, Eleanor, my wife, would call and ask, "The skies are clear. Are you staying to shoot stars?" Joy, when the skies were cloudy, I could come home for dinner with the family. EB workers made great progress, so the boat was launched on May 22, 1965. It was a festive occasion, the boat was clad in a special cover on her bow, the band played, and the ship's horn blared as the last blocks were released and *Polk* slid down the ways. The officers and crew were topside for what was an interesting ride. Our three children, along with others, escorted by Electric Boat personnel, were stationed under the bow as the blocks were pounded out. At this point in their young lives, they might not have known that Abraham Lincoln was the sixteenth president, but they certainly knew James K. Polk was the eleventh president of the United States. *Polk* successfully completed her sea trials in March 1966. *The New London Day* reported that Vice Admiral Rickover expressed his satisfaction with the boat. With all systems "go" *Polk* was commissioned on April 16, 1966, at the New London Underwater Sound Lab. The Blue crew departed on our first patrol in September. The boat was ultimately "homeported" in Rota, Spain, where a change of command was held between the Blue and Gold crews. I made three patrols before being relieved in December 1967.

The SSBN deterrent patrol pin is awarded for one patrol; each subsequent patrol is identified by a silver star. Achieving five patrols rates a gold star.

Polk's ship patch was designed by Captain McMullen's wife, Ruth. She was an artist and a very talented person. The significance of the emblem was explained in this way:

> *The burst of sun in the western sky and the eagle in flight symbolize the "Spread Eagle Platform" on which James K. Polk ran for and was elected to the presidency in 1844. He sought and achieved territorial expansion to the western borders of the United States. The four stars represent the four major land areas acquired during Polk's term of office: California, New Mexico and Oregon Territory, and Texas statehood... The blue field edged in gold reflects the founding of the Naval Academy in 1845.*

Building and commissioning the *Polk* made for a very cohesive Wardroom. As Navigator, I was the "third" officer behind the commanding officer and the executive officer. As such, I was the more accessible of the "senior" officers to the "junior" officers, eliminating the need for formality as shown by the special celebration for Mother's Day during our second patrol. Patrols were sixty-day affairs,

and humor was a good way to relieve the stress. Birthdays were celebrated as were Christmas, Mid-patrol, and other holidays. So, why not Mother's Day?

Polk had a successful career as a ballistic missile submarine, completing her sixty-sixth strategic deterrent patrol in 1991. She had twenty-five years of commissioned service at that time. In 1992, *Polk* began a shipyard conversion to an attack submarine. A dry deck shelter was installed on her deck which allowed her to support special warfare operations. The conversion was completed in 1994 and her hull classification was changed to SSN-645. She was now an attack submarine. She completed three extended deployments to the Med, taking part in numerous special forces exercises. A new patch was designed when she became an SSN.

James K. Polk was deactivated at Norfolk in January 1999, decommissioned in Bremerton on July 8, 1999, and stricken from the Naval Vessel Register. She was scrapped in 2000. Her sail is on display at the National Museum of Nuclear Science and History in Albuquerque, New Mexico.

Shaking Up the World 91

I am a *James K. Polk* plank owner. This means I was a member of the crew when she was commissioned. Members of the commissioning crew received a small model of the boat with the brass plate suitably engraved.

Traditionally, when a ship is decommissioned, the plank owner is entitled to a ship's "plank." When *Polk* was scrapped, various pieces were salvaged, and a special memento was created. The ship model shows the *Polk's* configuration with the dry deck shelter behind the sail. On the left is a piece of the hull and on the right is a link from the boat's anchor chain.

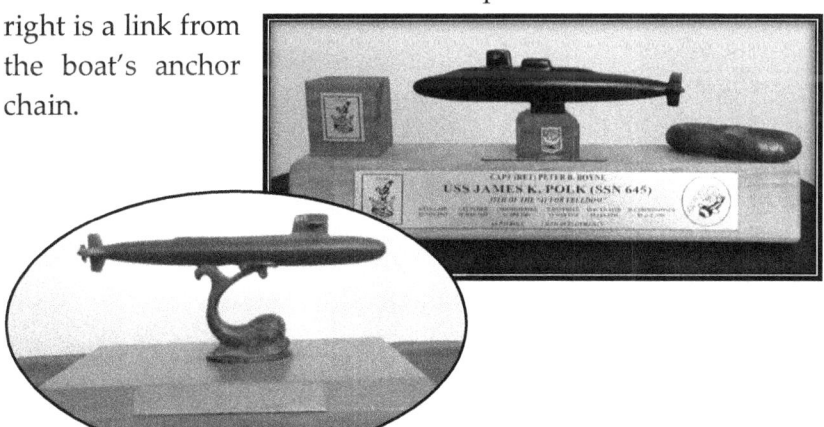

COUNTERINSURGENCY AND SURVIVAL TRAINING
Ray Dove

WHEN I RECEIVED ORDERS TO VIETNAM, it included a requirement to attend a "Counter Insurgency and Survival" training course at the Little Creek Amphibious Base. As a thirty-four-year-old Lieutenant Commander, I joined a group of twenty-five eighteen- to twenty-five-year-olds. There was one young officer, and one Chief Petty Officer, and the others were a mixture of enlisted men.

The first two weeks of the three-week course were classroom work, calisthenics, obstacle courses, and running. (A two-week Boot Camp but no formation marching or manual of arms.) On the third week, we were transported to Camp Hill, Virginia, and dropped off in the middle of nowhere. We were directed to camp out and live off the land for two days. They did provide us with one dead and dressed chicken as provisions.

It was the end of October, other classes had preceded us in this area, wild game had been all but obliterated and vegetation had succumbed to the elements. In our pursuit of edibles, we captured a stray dog, but it had a collar. A Filipino with us offered to cook it if someone would kill it. Instead, I attempted to make a deal with the instructors. We wouldn't kill and eat it and would turn it over to them for a case of C-rations. No deal, so we released it, after all, it was an awfully small dog.

On the third day, the training exercise commenced. Initially, we were told to spread out and hide ourselves in any manner we could imagine and to avoid capture. We had two hours, then patrols would be sent out to find us. All of us were found and rounded up by late afternoon. That evening we were given compasses and instructions to travel and to contact "partisans" by the next morning. "If successful, that would end the exercise." The instructors departed.

After trampling through the night, we all made contact. The instructors told the truth, that exercise was over, but a new one had commenced. The "partisans" were NOT friendly. We were stripped down to our T-shirts and underwear and force-marched to a stockade where a VC flag was flying on a flagpole. (Sort of a Bataan Death march, sans death or blood.)

On arrival, we were forced to our hands and knees and made to crawl into the stockade through muddy trenches. During this time, we were addressed as pig, swine, hog, and other epithets. Inside the stockade, we were subjected to humiliating and harassing incidents. Finally, a stocky "partisan" dressed in the uniform of a Green Beret came at me as if he wanted to kill me, screaming, "Pig, you're being uncooperative." I had been recognized as the senior prisoner present as a result of the minor resistance the partisans met. He stiff-armed me in the chest, grabbed me, dragged me, and bodily threw me into a three-foot deep, grave-shaped, pit. Then he and other partisans threw buckets of water on me, all the time screaming epithets and obscenities at me. The Green Beret fished me from the pit, ripping my T-shirt off (it may have been accidental, but I lost the shirt in the process), and continued screaming obscenities at me.

I was then dragged out of the stockade and thrown into an

isolation shack. The shack was a small structure, with no windows and the interior walls were padded with wrestling mats. There, he kept bouncing me off the walls, continuing with his screaming. I began to wonder, *Is this part of the training or sadistic pleasure for this guy? Are they going too far with the game?* I finally relaxed and decided to let them play their game. At this point, the Green Beret stopped the harassment and went out of the shack slamming and locking the door. I was left in complete darkness, with no windows, and no furniture. I sat down in isolation, shivering from cold and emotions.

When the shack door opened and I was ordered out, the sun had set, and it was dark. I was marched into the stockade and ordered by the partisan to bring my "Pigs to attention and honor the camp's evening flag-lowering ceremony." He pointed to where I assumed the flagpole was located. The stockade was dimly lit. I ordered, "Fall in, two ranks facing me." The prisoners complied, and I ordered, "Parade rest." As I turned to face the flagpole, a bugle sounded, and a powerful spotlight illuminated the top of the flagpole. A tingling sensation of patriotism went up my spine, there was no VC flag—instead, waving in the wind were the Stars and Stripes of America. I ordered "Attention, right-hand, salute."

It may sound corny today, but at that moment in time when everyone expected to see a VC flag, it was a dramatic and powerful conclusion to a day of Hell and three weeks of training.

EISENHOWER AND NIXON EXPERIENCE
Bill Smollen

IT WAS JULY 4, 1948, in Vicksburg, Mississippi, my hometown, which had not celebrated the Fourth since 1863, when the city surrendered to the Union Army in our Civil War. When Union General US Grant accepted the surrender from Confederate General John C. Pemberton (ironically from Philadelphia, Pennsylvania), it marked the end of the Confederacy. The Union Army would now control the Mississippi River, and food and munitions for Lee's army would no longer flow to Virginia from the west by rail.

There was, at that time in Vicksburg, a family named Lum with a lifelong friend who had become famous but wanted to escape that fame for a couple of weeks. Those friends were coming to visit with the Lums over the Fourth. As it turned out, the Lums told one person too many about their house guests. Then the Chamber of Commerce got into the act and decided that Vicksburg would celebrate that Fourth in honor of that particular house guest who happened to be Dwight David Eisenhower (Ike), a five-star General, the Supreme Commander of Allied Forces in Europe, and the architect of the WWII victory there.

My dad got me up quite early on the morning of the Fourth and made me put on my Boy Scout uniform. As compensation for that, he took me for a seaplane ride over the Mississippi. On our return, I inadvertently grabbed the hot engine exhaust to maintain my balance while leaving the plane, burned my hand badly, and had to go to the hospital for treatment. Because of that, we missed the early part of the parade. Eventually, we wound up at the courthouse just as the parade finished, whereupon Ike gave a speech that capped off the festivities.

We hung around as the crowd dispersed until General Eisenhower was unoccupied so we could shake his hand. To our surprise, when we were about to be the last folks there, instead of a quick handshake and goodbye, Ike was kind enough to spend about half an hour talking with me and asking me things to put me at ease. He asked how I was doing in school, how I was doing in Scouting, etc. What a neat man! I could see how he could get along with a tough coalition of foreign leaders.

It was a great, heady day for a thirteen-year-old.

Flash forward to January 1957, when I was a First Classman (Senior) at the US Naval Academy, and Ike had been reelected to the presidency. The Academy was chosen to march in his Second Inaugural Parade. Afterward, in the evening, there were several inaugural balls to which some of us had obtained tickets. I was dancing with my date, lost in the music and euphoria of the occasion when a man tapped my shoulder. We stopped dancing, I turned around, and this older man told us he was a Secret Service Agent. He said Vice President Nixon would like to speak to us. Suffice it to say I was completely surprised.

We went to the VIP area and were introduced and glad-handed all around. We met Vice President and Pat Nixon, their daughters, and some other folks. Everybody was having a grand old time, and these VIPs included us in their celebrations.

Vice President Nixon asked me where I was from, etc., etc., and did his best to put us at ease. After a bit, the VP asked me if I knew "undisclosed" (a classmate).

"Yes, sir! He's a good friend of mine! He's a good guy!"

He asked me if I could do him a favor, to which I quickly agreed and then he told me he had appointed "undisclosed" (the classmate) to the Academy and that he was getting a bit inebriated. Furthermore, VP Nixon was concerned that "undisclosed" might get into some trouble and embarrass both himself and the Veep before the night was out.

VP Nixon asked me to keep an eye on him and to try to keep him out of trouble. I readily agreed, and my date and I found "undisclosed" and casually checked on him (VP was right). We danced nearby and watched him out of the corner of our eyes. We did not allow that to detract us from having a great time, regardless.

So, I've done something I bet you have never done. I hustled a drunk for Richard Nixon!

Later that same evening, Joan and I were dancing away, still having a great time, when another Secret Service man tapped me on the shoulder and said the President would like to see us. As we worked our way toward the President, I told the VP we were on the job. When we were presented to the President, Ike led off with, "Bill, good to see you again. Did you ever make Eagle Scout?" Wow! I floated on that one for a month!

A GREEK HOLIDAY
Ted Kramer

THE TURKISH INVASION of the Greek portion of Cyprus in 1974 was further from my mind when the ship I commanded, the USS *Richard E. Byrd* (DDG-23), was directed to visit Corfu, Greece, for a three-day rest and recreation visit almost a year after the Turkish invasion. It hadn't occurred to me either that the United States had supported Turkey in their invasion despite their atrocities and deprecations of the Greeks on the island. I had visited Corfu shortly before the Turkish invasion on a different ship and was looking forward to a return visit to this lovely tourist haven.

Our anchorage was close enough to Corfu's only large pier that would accommodate a destroyer, but it was occupied by what looked like a large ferry with a group of women on the pier next to the boat, some waving goodbye, some weeping. Per our normal protocol, I sent an officer in to arrange tours for the crew, and I also sent our Disbursing Officer and his clerk into town to exchange dollars for Greek drachma. By then, a Greek Navy Lieutenant Liaison Officer (GLO) arrived with instructions. He spoke very good English but seemed extremely nervous. He stated that the Mayor and Chief of Police wanted to see me in the mayor's office as soon as possible. I asked the Lieutenant what was going on at the pier and he told me a bunch of conscripts were being loaded on the ferry for a little training before being shipped off to Cyprus to fight the Turks. He asked that we not send any liberty parties ashore until after a demonstration by the Communist-led Taxicab union was done in the early afternoon because the temperature was still high against the American support for Turkey. "Fine," I said, then loaded my gig with my Chief Engineer and Weapons Officer (safety in numbers) along with the GLO, and off we went.

The plan was to moor the gig on the other side of the pier the ferry occupied, but the GLO (still nervous) didn't think that was a good idea as the taxicab demonstration was approaching the land side of the pier and it looked menacing and decidedly anti-American, to say the least. Although the many banners and flags the demonstrators carried were in Greek, the pictures of Uncle Sam, the American Eagle symbol, the American flag, and other unmistakable American symbols being derisively characterized as anti-Greek corroborated the GLO's comment about the temperature in the town. The GLO, probably worried he would be caught in the demonstration and possibly identified as being on the wrong side, asked me if I would take the colors down from the gigs after Flagstaff. I replied that taking the American flag down would not fool anyone anyway, so we left the colors flying which prompted the GLO to get on his walkie-talkie for some animated conversation with someone ashore. When he was finished, he directed us away from the ferry landing and to the commercial merchant piers a short distance away. It was sealed off from the public by high fences topped with razor wire, a good safe haven. Three police cars were waiting for us, one for us and two escorts to the mayor's office.

The scene was a little chaotic at the mayor's office. After a few short formalities, the Mayor sat down at his desk and began communicating on a red phone he pulled out of his desk drawer. The GLO whispered to me that it was a direct line to someone in Athens. He made several red phone calls to Athens. Finally, a messenger rushed in and said something to the Mayor, who breathed a sigh of relief and picked up the red phone for a last call to Athens. The tension in the room broke and things became much more cordial. The most difficult part of all this was I had no idea what everything I just observed was all about but began worrying about the safety of the three men I had sent ashore earlier. I could only revert to the ancient idiom: "It's all Greek to me."

Whatever the crisis was, it appeared to be significantly over. We were then driven back to the commercial area where the gig was patiently waiting for us and took us back to the ship. Once there, I learned what all the red phoning and hullabaloo were all about. Our Disbursing Officer and his clerk were seriously threatened by the demonstrators along with the irate ladies seeing their loved ones off to Cyprus to fight the Turks. Our men ran down the pier, being pelted by rocks and garbage, and took refuge in a Hertz rent-a-car we had rented for our visit. That didn't help a great deal because the Taxicab Unionizers set the car on fire and turned it over. Our two men managed to get out of the car before any damage was done to them, raced down to the end of the pier, and jumped into the ship's whaleboat that was motored up to the pier for them to take back to the ship. When I saw Lieutenant Owen, the Disbursing Officer, after I returned to the ship about an hour after the two men had returned, he was still holding his briefcase full of money in a death grip and was considerably and understandably shaken by the incident. Somehow, the Tour Officer, while sipping ouzo at an outside café, told us he was delighted to witness an interesting and colorful parade march down to the pier. Right! He later managed to catch a water taxi back to the ship without incident and was still innocent of the gravity of the situation.

I gathered as many facts as I could and sent the obligatory "Pinnacle Message" to the proper higher authorities, set security details topside, and had my ship's boats manned with armed sailors in case the taxicab union had any further ideas other than destroying a rental car. (Hertz sent us a bill for the car two weeks later which we gratuitously ignored.)

Commander Sixth Fleet replied to my "Pinnacle" by telling me to weigh anchor and leave immediately. I replied to that message saying the Greek Navy felt all of this would blow over by the morning and that Corfu would be back to its normal touristy state and it would be safe to grant liberty for the crew.

Sixth Fleet replied to this request by saying, "I SAID LEAVE NOW!" Again, tempting fate and perhaps a career, I asked if we could at least leave after dark to save a little face. Commander Sixth Fleet, I could tell, was getting a little annoyed—although I couldn't read between the lines because there was only a one-word message: "LEAVE!!!" So, we left in broad daylight with our tails between our legs and there must have been several thousand Corfu citizens lining the shore waving goodbye to us with their middle fingers. I had the crew inside the ship just in case anyone on the shore wanted to say goodbye with a trigger finger vice a middle finger. My Weapons Officer kept prompting me that one round of "Willy Peter" (White Phosphorous) would take care of everything. Thanks, but not at the moment, Weps.

POST CORFU:

We took a leisurely trip across the Adriatic and landed in Ancona, Italy, early the next morning. I was able to get on the phone with Commander Cruiser-Destroyer Force Atlantic in Norfolk to give him the straight story, which was quite a bit different from the national news. Walter Cronkite on CBS had me "lynched and the three men I had ashore taken hostage." Dottie, my wife, was taking phone calls from family, friends, members' wives, the press, etc. I was able to get through to her on the phone and she was not only glad to hear my un-lynched voice, but to know everyone on the ship was safe. My twelve-year-old daughter got on the phone and wanted to know why I let some of my crew go ashore to smoke marijuana. I asked her what in the world she was talking about, and she said that Walter Cronkite said on the news that the officer and his clerk had run down the pier and were stoned. I explained the difference between rocks and stones.

We had planned to fuel in Corfu, but obviously couldn't, so we arrived in Italy amid one of their ubiquitous fueling strikes throughout the country. We were down to 15 percent of fuel in our tanks, which isn't good for the perennial thirsty Adams class Destroyer, so Sixth Fleet was gracious enough to get a tanker underway on our behalf. I was expecting our lights to go out and the ship to stop with no fuel any minute by the time we rendezvoused with the tanker, but we made it and thus ended our Corfu wonderful Greek holiday.

CHARLIE DUKE, ASTRONAUT
Charlie Duke

I FELL IN LOVE with airplanes during my junior year at Annapolis and decided to go into the Air Force. After graduation from the Naval Academy, I was assigned to Spence Air Force Base near Moultrie, Georgia, for pilot training, and then to Webb Air Force Base in Big Springs, Texas, where I earned my wings in September 1958. Advanced training was at Moody Air Force Base back in Georgia again to fly F-86Ls. My first assignment was to the 526th Fighter Interceptor Squadron at Ramstein, Germany. I reported to the 526th in May 1959. In 1961, Yuri Gagarin

was the first human in space, followed by Alan Shepard as the first American in space. Then President Kennedy announced we would land on the moon by the end of 1969.

My three years at Ramstein were finished in May 1962, and the Air Force sent me to MIT for a master's degree. While there, I was working on the Apollo guidance and navigation system MIT had the contract to build, and I had the pleasure of meeting some astronauts who were very motivated with their jobs. I asked them how I could become an astronaut. They gave me the advice to go to test pilot school, which could give me a chance to get into the program.

While I was at MIT, I met the love of my life, Dotty Claiborne, and we were married in June 1963, and now we are celebrating sixty years together.

After graduating from MIT in 1964, I reported to the USAF Test Pilot School at Edwards Air Force Base in Southern California and, after graduating a year later, I stayed on as a member of the staff, and our eldest son was born there in March 1965.

In September 1965, NASA had another call for astronauts, so I applied for the program and was selected with eighteen others in April of 1966. We welcomed our son, Tom, in 1967, and we were selected to fly in the Apollo program. I thought my chances were slim to none, but we had eight astronauts killed over two years. Three were killed in a fire on Apollo 1, then four were killed in airplane accidents, and one was killed in an automobile accident. Some senior astronauts retired, and our group started to get flight assignments. My first assignment was to the support crew on Apollo 10. Apollo 10 was the first mission to carry the lunar module to the moon and to simulate a landing without landing. I was the CapCon (Capsule Communicator) which in Mission Control is the person that talks to the crew. The mission was designed to circle the moon and return to Earth.

As a result of my Apollo 10 experience, Neil Armstrong invited me to be CapCon for their actual landing attempt. The descent was far from trouble-free. First, we had communication problems, then guidance computer overloads, and then a trajectory error that resulted in a minimum fuel problem. The crew landed with less than thirty seconds of fuel remaining. Tensions in Mission Control were through the roof! The rest is history, and Neil Armstrong stepped down from the lunar lander onto the surface of the moon, the first person to do so.

My next mission was as a backup lunar module pilot on Apollo 13. As we were approaching the launch of Apollo 13, I caught measles a week before launch, which created a big problem as noted in the movie, *Apollo 13*. My proximity to the crew exposed all of them and consequently, Ken Mattingly was removed from the crew for fear he would come down with measles. Rusty Swigert from the backup crew took his place. Sorry, Ken!

After the Apollo oxygen tank explosion, Mattingly went to Mission Control and stayed there for thirty-six hours to help devise a plan to get the Apollo 13 crew back down safely. The expertise of Mattingly and Mission Control saved the lives of the crew, no question. They made it home safely. John Young and I did our bit to help recover the crew.

John Young, Ken Mattingly, and I trained for the Apollo 16 mission which was delayed because the mission was to be for seventy-two hours on the surface of the moon instead of twenty-four. A rover would be carried for our use to explore areas farther away from the lunar lander.

Happily, I was selected as the lunar module pilot for Apollo 16, and as we sat on the launch pad at Cape Canaveral, we had a four-hour window when we had to launch, or it would be another thirty days before the moon floated into a position for us to launch again. Launch procedures went well thankfully, and we lifted off on schedule.

The Saturn rocket was a little shaky for us, but we got into orbit as designed, and our view was spectacular! As I looked out the window, the colors of Earth were something to behold. We could see the blue of the ocean, the white of the snow and clouds, and

the brown of the land, grass fires in Africa, and a big line of thunderstorms.

Our trip to the moon was three days with no problems on our way out. After our arrival on the moon, we spent one day in orbit getting everything ready. From the moon, I must mention the earth was a beautiful jewel of blue, white, and brown suspended in the blackness of space. On April 20, 1972, John Young and I put on our space suits and in two hours, we powered up the lunar lander and floated free. One hour before we were to land, there was an issue that almost caused us to abort, but Mission Control went to work to find us a workaround procedure and a landing spot. Though we were now six hours behind schedule with new procedures, we started our descent, and John Young did a great job of putting us down in a good spot on the moon where we would spend almost seventy-two hours on the surface. Our excitement level was at the max as we looked at the mostly gray, rough, barren surface nobody had ever touched before. Mission Control changed the flight plan and told us to go to sleep for four hours. That was crazy! As excited as we were, sleep was not easy, but we did what we were told to prepare for the days ahead. What a ride that was!

During the next three days, we made three excursions in the lunar rover we deployed soon after landing. It took us two to three

hours to suit up for these extrusion trips with all the gear needed. The shades of gray made the lunar surface magnificently beautiful, but it was a hostile environment where everything had to work, or we were dead. On our second excursion, we climbed a nearby mountain in the rover three hundred feet above the surrounding area, and we could see our module from there. There was constant excitement the entire time, we didn't know what we would see over the next ridge. John drove, and I navigated, using a travel guide from Mission Control. The terrain was rough, and we bounced around. I was describing what we were seeing to Mission Control. The rover would do seventeen kilometers per hour, and with the roughness of the ride and fishtailing as if on ice, we wore seat belts. We were limited to a four-mile radius from our module because, in our training, we learned we could only walk that far in our space suits. Fortunately, our car worked well.

With the Summer Olympics coming up on Earth, we decided we should do a couple of events on the moon, a high jump, and a long jump—a risky idea looking back. I started to bounce for a high jump and was falling on my back, so I rolled to my right and broke my fall before landing on my back—the important backpack was not damaged. At least I was still alive. John looked at me, "Not very smart, Charlie!"

"Help me up!"

Mission Control piped in, "No More."

The games were over. We decided more of this kind of fun would have to wait until we got home. However, John and I did set a high jump record on the moon which may never be broken. Twenty years later, the Olympic committee presented us with a trophy for taking the spirit of the Olympics to the moon, and although it wasn't a gold medal, it was a very special trophy for us.

I had permission to take two items to the moon. One was a family photo we left there after taking a photo of it on the moon. The other item was two Air Force twenty-fifth Anniversary Medallions—one we left there, and the other is at Wright Patterson Air Force Base in Ohio. Happy birthday, Air Force!

We packed up the rock samples we collected, left the rover and backpacks on the moon, and headed for Ken Mattingly in the orbiter. When we got there and engaged properly, Ken opened the hatch and would not let us in his spacecraft because of all the lunar dust floating around our lunar module. The dust had come into the lander on our boots and was floating around everywhere.

"You are not coming in here with all that dust, get cleaned up, and I'll let you aboard," Ken said. He handed us a vacuum cleaner, and we went to work to clean up the dust and transfer the rock samples.

What an adventure! We didn't want it to be over, it was such a sense of wonder to me, a very humbling experience. No one has explored our landing area in all of history, and I will be forever thankful for the opportunity. We made it home safely and accomplished our mission. Our country made nine trips to the moon and landed there six times, and 10 percent don't believe it. That's unbelievable. If you are gonna fake a landing, why fake it six times?

I left NASA in early 1976 to enter private business and finished my Air Force career in the Reserves, working at Recruiting Service where I was promoted to Brigadier General, retiring in 1986.

Since 1978, I have been involved in various businesses and motivational speaking, and my wife, Dotty, and I have been active in Christian ministry which has taken us all over the world to share God's Love. Well, that's my story, and I'm sticking to it. Hope you enjoyed it. I was so proud to represent our Class of 1957.

ENGINEER TO NEWSPAPER OWNER
Gerry Anderson

AFTER TWENTY YEARS as an Air Force officer and sixteen years working in the defense industry as an engineer, it was a total shock for my wife, Edie, and me to find ourselves owners of the weekly newspaper in Molokai, Hawaii, *the Molokai Dispatch*.

A sequence of events led to this unexpected happening. In 1987, when I was transferred to the newly opened Hawaii office of my major corporation in Kailua, Oahu, we tried to purchase a home on Oahu, but everything had some problems, including being overpriced, leasehold land, being poorly maintained, or spotty neighborhood. Thus, we made the fateful decision to purchase a home on the neighboring island of Molokai—with five acres, and a spectacular view over the channel to Diamond Head.

While I commuted to work daily via a small commuter airline (twenty-five dollars for a round trip then), Edie started writing for a bimonthly newspaper *the Molokai Dispatch*. In 1992, the newspaper changed hands, became weekly, and we became more involved—with Edie writing more and me joining the Board of

Directors since the owner was a resident of the mainland and at least one resident was required to be on the Board.

In July 1993, after returning from a European cruise, we were informed by the owner that he was moving to Texas, and we could either purchase the paper for the cost of the equipment, or it was going to close. Since we did not want to see Molokai without a newspaper, we took over.

We found that owning a business in a small rural island community like Molokai presented many challenges. The population of Molokai was less than seven thousand, with many on welfare, and drug problems were rampant. The schools were terrible and families that cared about education moved away. Of the sharp kids who finished high school on Molokai, most moved away after graduation, never to return. This left the remaining population with few skills and no work ethic; we could never plan on an employee showing up on Monday morning. There were very few facilities on the island, no supermarket, and, in our neighborhood, no cable TV. We had to have the newspaper printed on Oahu and flown back.

When we took over the paper in 1993, we found it to be a money-losing operation with a bloated staff. We immediately reduced the staff to two, and I attempted to run the paper while continuing my full-time corporate job. Early in 1994, we found that one of our employees was embezzling money from us, and she eventually served six months in prison and had to pay partial restitution.

At that point, we were down to one employee, and I reduced my corporate job to part-time, retiring completely in 1997. Edie and I rapidly learned everything to put out the weekly newspaper, covering events, writing, selling advertising, page layout, subscriptions, and bookkeeping. We did it all ourselves because of the lack of reliable employees on Molokai, and it required seven days a week of effort.

The only way to survive this high-pressure environment was to totally get away from Molokai two or three times a year. The most enjoyable way was to go on cruises, which we did regularly starting in about 1995. The best part of cruising was the total isolation, and if something went wrong with the paper, we did not find out about it until we got home. The bad parts were that we had to prepare the issues while we were gone, leaving provisions for last-minute ads. This meant we had to find someone reliable to do the required last-minute things to the paper, get it to the printer on Oahu, and make sure it got distributed, and we always had to correct the screw-ups that happened while we were gone. But getting away was worth it. We also wrote up many of the places we visited, which turned out to be a very popular travel series.

The good parts of running the newspaper included meeting interesting people and attending interesting events, including the fiftieth-anniversary celebration of World War II on Oahu, with President Clinton attending the dedication of the Battleship Missouri museum in Pearl Harbor.

One amusing incident occurred in Washington, DC, when Edie had a White House Press Pass and was in the press room. Suddenly, a group of reporters left to go elsewhere in the White House, and Edie joined them. It turned out to be a small meeting between Hillary Clinton and the press to discuss healthcare ideas. Everyone from the press was asked to introduce themselves. Along with the major newspapers, was "Edie Anderson, *Molokai Dispatch* in Hawaii," to which Mrs. Clinton responded with a comment about the fine health-care programs in Hawaii, and Edie corrected her, "It's not very good."

In 2003, we started reevaluating our priorities and started thinking about selling the paper and getting out of the rat race. We finally did in March 2006, selling to a thirty-year-old former Molokai boy who moved back from Montana. We are now finally retired, and *the Molokai Dispatch* is under a new generation.

One thing we can say with sincerity is that through the years, USNA and the Class of 1957 have come to mean more and more to us, and classmates have become extended family and remind us of what remarkable people make up our alumni. Mahalo Nui Loa!

A SOVIET BLACK SEA CRUISE
Sam Coulbourn

BLACK SEA CRUISE WITH THE HAMMS (FEB. 1983). It was February in Moscow, which was cold and rather miserable. My wife, Marty, and I were assigned to travel down to the Black Sea with our General and his wife, Jane, for a week's cruise. As always, this was an intelligence collection operation—this time to photograph and observe warships in the Black Sea ports Odessa Batumi, and Sochi, and to observe whatever else.

My boss was Air Force Brigadier General Charlie Hamm, a former Thunderbird pilot.

When the KGB heard an American general was taking this cruise, there must have been a scene down at headquarters, lining up all the agents who were due for a week in the warm climate, and picking out the likely candidates.

The flight from Vnukovo Airport in Moscow to Odessa was typical Soviet. As soon as you stepped into the cabin, you could smell the delightful aroma of unwashed armpits, hydraulic oil, stale bread, garlic, and vents from toilet tanks with filters that hadn't been changed. Marty sat in a seat, and it was wet. She pointed this out to the stewardess, a muscular big bleach blond. The stewardess simply reached down and grabbed a little man in the seat ahead of Marty, pulled him up by his coat and, at the same time she asked Marty to move aside, plunked him down in the wet seat. Then she ordered Marty into the dry seat.

Aeroflot IL-86

On Aeroflot, if you're flying on Election Day, Soviet citizens vote in-flight.

And the food—the food on Aeroflot was unique. Sausages in heavy cellophane casings, served with lumpy mashed potatoes and green peas as hard as marbles. But the butter (each pat stamped with a hammer and sickle) was good.

The stewardess delivered a speech before we arrived at each city, about its "hero" status, how they fought valiantly in the Great Patriotic War, and how many factories, theaters, and schools it has. On the flight to Odessa, we flew over Kiev, and they even gave us a speech on that hero city.

When we boarded our Soviet cruise ship, we counted twenty-four agents of all sizes, shapes, and sexes, on the cruise ship. They even had a male and female agent team posing as a newlywed couple on their honeymoon. Our first clue as to their real status was when the groom kept pretending he was shooting photographs of his beloved bride, but shooting us, instead.

Churchill, FDR, and Stalin at Yalta, February 1945.

We landed at Yevpatoriya, Ukraine, on the Crimean Peninsula. Breakfast on the Soviet cruise ship was macaroni and meat sauce, cheesecakes with sour cream, grape jam, cheese, and bread. An Intourist Guide took us on a fast, picturesque drive along the peninsula to Livadia, where the Czars went to get away from the frozen city of St. Petersburg.

We toured Livadia Palace, where the Yalta conference was held with Franklin Delano Roosevelt, Winston Churchill, and Josef Stalin in February 1945, just two months before FDR died. We saw drab little old men in dark, rumpled suits with medals, and large women with fluffy sweaters touring the palace.

One thing we discovered was that the farther from Moscow you got, the more the KGB resembled Keystone Kops.

On our cruise, we visited several Black Sea ports, and we'd go ashore and see the sights. The agents would follow us or be stationed to watch us before we went ashore. The obvious head of our surveillance, whom we called "The Main Man," was always standing somewhere, watching us. In Yevpatoriya, we passed under this large old boat up on blocks. As we looked up at the boat above us, there was a little agent up there, busily photographing us.

Some agents would follow us, dodging from palm tree to palm tree behind us. Then, just as they must have been taught in spy school, they'd go swap disguises. You'd see these men switching caps or wigs, or changing coats when they thought we couldn't see.

On the ship each night, we'd have dinner together and then go to the lounge and have a Soviet brandy and listen to the orchestra play, and maybe dance. One night, Jane suggested to Marty we skip the lounge scene and just get to bed early.

We knew microphones were listening in on us all the time.

As it turned out, after dinner, the ladies changed their minds and suggested we go to the lounge for a nightcap.

This was a surprise for the KGB. Usually, there was an orchestra playing away for us, and for all the agents who were masquerading as happy passengers.

There was no one but us in the lounge. Soon a waiter appeared to take our order.

Then, a few moments later, looking as if they had been roused out of a sound sleep, came the orchestra playing. Then, a few minutes later, came the "passengers."

Sam Coulbourn, 21st Co. started in destroyers, then went to submarine school. He served in a diesel submarine and then was the weapons officer aboard a Polaris submarine, making four patrols. He went on to serve as navigator aboard the National Command Post Afloat, then commanded USS *McCaffery* (DD-860) in the last months of the war in Vietnam, and commanded USS *Mount Baker* (AE-34) during the Yom Kippur War in the Mediterranean. He served as naval attaché in Moscow for two years, then commanded a naval base in Sasebo, Japan. After thirty years, Sam ran a leadership training course at Endicott College in Beverly, Massachusetts. He and Marty settled in Rockport, Massachusetts. She died in 2018.

FINDING THRESHER
Jim Gradeless, TM 1, Friend of USNA Class of 1957

SIXTY YEARS AGO marked a time in my life I shall never forget. I was a crew member on the submarine USS *Redfin* (SS-272), and I had just got home in Norfolk from the boat when the phone rang. The call was a recall of the crew. I reported back aboard the *Redfin*, and we got underway on the eleventh of April 1963, with a skeleton crew of less than fifty hands.

We were headed to a point 220 miles east of Cape Cod where the USS *Thresher* (SSN-593) had just sunk. This boat, the *Redfin*, was selected to be a part of the search team for the *Thresher's* resting place for a specific reason. The *Redfin* had an extra thirty-foot compartment added to the boat some years earlier, and this compartment now contained a lot of advanced-sounding gear not found on other boats that would later be deployed to the new nuclear power subs.

We searched for flotsam, anything that may have been a part of the *Thresher*, and we conducted a grid search as thoroughly as possible. I am led to believe the *Redfin* pinpointed the wreckage of the *Thresher* at 9,400 feet below the surface, but that information is for those above my pay grade. After locating the *Thresher* and marking the location with the latitude/longitude, the ships on station there conducted a memorial service.

The entire time on station, the weather had been cold, and the seas were rough and choppy. Now, on April 15, at 1715 hours (5:15 p.m.), sailors not on watch manned the rails in their dress canvas. We had become a part of a procession of ships lined up at 1,500-yard intervals. The seas had become calm and were almost a slick surface. It was drawing near sunset as we passed the location where the *Thresher* was resting. God bless her crew and a few civilian shipyard workers that were on board at the time of going down, 129 total. As the procession of ships, USS *Warrington* (DD-834), USS *Thomas Jefferson* (SSBN-618), USS *Sunbird* (ASR-15), USS *Redfin* (SS-272), and the USS *Rockville* (PCR-851), went by the marked location, all hands manning the rails rendered a hand salute while the USS *Warrington* fired a twenty-one-gun salute.

At this time, a Naval aircraft flew over and dropped a ton of roses upon the location. All hands like mine had tears streaming down our faces in the cold breeze. On April 21, 1963, we completed our mission and sailed back to Norfolk, where we moored at the D&S (Destroyer & Submarine) piers and tied up at Pier 21. Later, I was selected to be on an Honor Guard at the Naval Academy Cemetery at Annapolis MD. where a memorial service was conducted for the Skipper of the *Thresher*, Lieutenant-Commander Harvey. I am proud I was there for the sailors who gave their lives and could be a small part of Naval history. I will carry these memories to my grave.

GLASGOW MARINE COURT
Mike Giambatistta

"MIKE, PREPARE TO BRIEF THE CREW about visiting Glasgow," directed the Executive Officer, Dave. The most succinct advice was contained in one of the several available documents: "Glasgow is a good place to visit if you want your front teeth knocked out."

Glasgow Marine Court

"Mike, I've got another job for you."

"Dave, it better be a quiet one, my head feels—as Miles Graham says—like a nickel watermelon," I said.

"What you did to yourself last night is your problem. Jim Blackburn, Red Dog, and Mike Lintner must appear in the Glasgow Marine Court at ten a.m. this morning! Seems our boys thought the taxi fare was exorbitant and decided to offer the driver a knuckle tip or two! Your job is to suit up, be there as a witness of the proceedings, and provide character references the court might require," directed my Executive Officer.

This was the proverbial "Message to Garcia" moment—where is the damn place, and how in the hell do you get there from here (which was Faslane, the Royal Navy submarine facility in Gareloch in the Argyle Highlands)? Bus, train, taxis, and lots of the Navy's all-purpose painkiller, "APC," did the trick!

On arrival, the court clerk patiently explained that the Sheriff would soon start the hearing. Scottish courts are presided over by a Sheriff, who is a legally qualified judge. Unlike our Wild West version, the Sheriff didn't wear a badge, carry a gun, or wear spurs.

It was hard to believe how shipshape and squared away the defendants looked as they appeared before the Sheriff. Red Dog seemed almost angelic as he guilelessly related his version of the difference of opinion with the Glasgow cabby, and subsequently the police, in his drawling Tennessee accent. "Bush" Blackburn stood in the dock half-deaf from time spent in submarine engine rooms (protective "earmuffs" hadn't been introduced yet). Anyway, the pounding in his sore head from the previous evening's festivities probably destroyed the fidelity of any sounds that might have filtered through. To further complicate matters, the thick Scots "Glaswegian" accents of the court officers added to the confusion. After almost every statement or question from the bench, Blackburn cupped his hands to his ears and shouted: "What? What did he say?"

After about five minutes of this pantomime, the Sheriff

shrugged his shoulders, pounded his gavel, and gave his decision—unique to Scots Law—"Not Proven."

The troops got the drift because they quickly disappeared to the nearest pub while I went up to the court clerk to clarify the status of our guys. "What does 'Not Proven' mean?"

"Get out of town," was the clerk's response.

Overhearing my question, the Sheriff said, "The interests of justice were best served by this decision. Malfeasance obviously occurred but the circumstances, the defendants' genuine answers, and the good conduct you attested to convinced me to close the issue with no assignment of guilt or innocence! That's what 'Not Proven' means."

GUARDIAN ANGELS
Ron Goldstone

MY INITIAL SEA DUTY assignment was Damage Control Assistant on *Stormes* (DD-780). During underway refueling, I was nominally in charge of the forward fueling station on the level just below the bridge.

One dark and starry night, I believe, during refresher training, we conducted refueling operations near Guantanamo Bay, Cuba. It was a balmy, Caribbean evening, and we were cruising at the usual twelve- to fifteen-knots speed, which provided a cooling effect on the skin—comfortable!

Stationed alongside the oiler ship, the hose was secured in the trunk, and we began pumping. I was positioned directly below the fuel line, executing my assigned duties flawlessly. Suddenly, the hose ruptured directly above my head and pressurized bunker fuel rained down on everybody in the vicinity.

I seem to recall that fuel oil was heated slightly to improve viscosity during pumping. Either that or the friction during transfer caused the oil to heat up.

Anyway, my skin immediately sensed a distinct temperature difference between the oil and the cooling effect due to relative wind passing over my exposed areas. My initial reaction was that I was being boiled in oil and my brain told me to escape! However, the slippery deck was unresponsive to my soles and luckily, my lifeline kept me from falling to a lower deck.

I often shudder when I think about what it might have been like in the water, at night, drenched in fuel oil, watching the oiler and my destroyer steaming away at twelve to fifteen knots. Thank goodness for guardian angels!

HOW I SPENT MY CHILDHOOD IN A WAR ZONE
Walt Meukow

I DOUBT MANY PACKED as many thrills, excitement, and danger into four years between the ages of nine and thirteen as I did.

I was born in Chefoo, China, in 1932. On December 4, 1941, my mother, stepbrother George, and I departed Shanghai on the *Marshal Jofre*, the last liner out of Shanghai before war erupted there, to join my stepfather (a retired Chief Quarter Master) in Cavite, Philippine Islands. We arrived in Manila late on the seventh.

When we finally hooked up with a somber Dad, we found out Pearl Harbor had been bombed—we were at war! Our first order of business was to pack two suitcases and be ready to run to the old Spanish fort, a half block away, to the USMC headquarters when the air raid siren was heard. This would be our bomb shelter.

During the Japanese bombing of the Cavite Navy Yard on the tenth, my mother, preparing her first cooking adventure without the help of a cook, was late. Dad, a very Navy-trained Petty Officer, took off—infuriated without lunch because he was late.

Immediately after he left, the air raid sirens went off. When Dad arrived at the Cavite Navy Yard gate, he was quickly ushered into a convenient bomb shelter there. When he finally reached his supply department office after the air raid, he found the building had received three hits. All his colleagues were dead. I never again heard him complain about Mom's cooking. Being late saved his life!

Although uninjured in the dense old Spanish fort, we lost virtually all our possessions when our newly rented house took a direct hit.

Following that, we took refuge in a small Filipino barrio on the shores of Manila Bay, awaiting the end of what we expected to be a short war. When the action picked up immediately after New Year's Day, 1942, we took the advice of the friendly Filipinos—to head for the sanctuary of Bataan, across the bay.

We left there in a heavily loaded Banca with three Filipino fishermen, four of our families, and an Army Air Corps pilot. Our voyage started with a huge surge from the river, and then we were becalmed for hours. When the wind picked up, we were finally underway again, and we experienced leisurely sailing; then increasing winds; then slight rain; then a storm with torrential rain. Bailing the boat was by hand, and not very effective. One of our Filipinos experienced a large gash across his face when a halyard line was blown loose and lashed his face.

The aftermath of the storm was a dense fog as we drifted without a sail. Suddenly, a dark shape appeared, and we heard the faint thumping of diesel engines. Dad commented that he wished he had a flashlight. Brother George piped up and offered a flashlight given to him just days before by the Army Coast Watchers at Timbaland.

A frantic SOS was answered by the USS *Finch* (AM-9). We were pulled on board seconds before our small Banca broke apart. After tucking the children in, Dad spent the night in the chiefs' quarters, checking out the latest grim situation. The Skipper had been Mom's tennis partner in Chefoo just months ago, though it seemed like years now.

Early the next morning, the *Finch's* Coxswain dumped us at Cabcaben pier despite the frantic signs of an Army MP waving us off. We were bombed minutes after landing. After the raid, we were attached to Field Hospital #2 for rations and an abandoned nipa shack for shelter. I visited the wounded and sick stretched out on the jungle floor in open-air wards, or flimsy tents, and played chess with them, until chased out by one of Elizabeth Norman's "WE BAND OF ANGELS" nurses.

The mess served us Wainright's cavalry horses, albeit each share was minuscule. Have you ever eaten horse meat? Routinely we heard Corregidor's "big guns" thunder over the hospital, and we knew we were still in the fight. One evening, we heard a hell of a rattling noise and a loud crashing noise about two hundred yards from the hospital. It was one of our last P-40s trying, unsuccessfully, to reach the dirt landing field near Cabcaben. I don't know if the pilot survived.

In March 1942, we were provided with a dilapidated school bus, missing all its left side, for shelter. It was better than our leaking nipa shack and was located now in a valley supported by a contingent of the 57 Philippine Scouts, an elite group.

On April 8th, the Japanese finally broke through our impoverished lines. With absolutely no idea as to where we hoped to escape, we drove our bus down the east road to Mariveles, passing some of our troops marching north and some marching south.

We drove through the burning fuel depots in Mariveles, which scorched the paint on our bus and made breathing difficult. We finally made it through to clean air and drove north on the west road. We met a contingent of Filipino Scouts who advised us we had surrendered and were to gather at Mariveles with other prisoners. We took the scouts on board and flew a white bed sheet identifying our intended destination. George and I dismantled quite a few weapons and flung them into the jungle bordering the road.

On the way, a trigger-happy Zero pilot strafed us. I guess just for fun, or because he was either inept or not serious. The bus received a few calling cards, but there was no real damage; other than having to suck up the six inches of dust covering the road in the dry season we all had dived into.

At the surrender camp, we saw hundreds of dejected and tired troops squatting in formation while the Filipino troops were separated and subjected to harsher treatment. There, we met US General King and a Japanese General and, after loading our bus with Filipino civilian refugees, we were directed by them to report to Manila. After leaving the surrender site, we passed a contingent of our troops in the infamous Bataan Death March and shared our rations with them.

In San Fernando, the site where the American and Filipino soldiers were put on trains to the infamous Fort Donaldson, an overzealous Captain stopped the bus, accused us of spying, and threatened to shoot my parents. Then he ushered us into an abandoned nipa shack.

The bus, with an inexperienced driver, was sent on its way with grinding gears and weaving down the highway to Manila. Shortly afterward, the Captain returned with a doctor in tow, having recognized that Dad was shaking with an obvious case of malaria. He then gave George and me a large tin can, took us by the hand, and led us to an open kitchen for an awful-smelling shrimp and rice meal that went down easily by our starving foursome.

That evening, he reappeared with a Tea Caddy and biscuits. The same on the next day, and on the third day, he showed us pictures of his family, whom he had not seen for years because of the China campaign. He started to cry. My mother put her arm over him and said: "This too will be over." I doubt such would be for him, but please think of his emotion at the height of the Japanese military surge. I for one have developed a very conciliatory feeling for the normal Japanese.

He then advised he was sending us to Manila, specifically to the infamous Fort Santiago, for interrogation. He said: "Tell the truth and all will be well."

Our "quarters" in Fort Santiago were fifteen-by-fifteen cells, devoid of anything other than a Filipino family of four. No pillows, chairs, NADO! "Facilities" consisted of a trough at the far end with constantly running water to move waste. Twice a day, a saucer of chocolate milk and half a loaf of bread was shoved through a hutch in the door.

George and I were let out each day for a half-hour romp in the large corridors, which was not so for the Filipino kids. Compared to the military POW camps and later the Hilton in Saigon, it was an uncomfortable cell, but there was no rough stuff on Dad; although we heard enough throughout the night to know there was different treatment going on.

After ten days, our family was released and sent to Ateneo de Manila, a Jesuit-run hospital, to recuperate before being trucked to Santo Tomas Internment Camp (STIC). Dad's condition kept us comfortably in Ateneo de Manila for the better part of a month because he had a strong case of malaria.

For those who were torn out of their lives of luxury and abundant social lives, and thrust into a restricted, crowded environment, and mundane diets, it was a tough change of pace. Families were split up, men mostly in the gym, women throughout the former classrooms or office spaces in an administration building. For our family, bombed out of all our possessions and surviving three and a half months in the mosquito-ridden jungle of Bataan, with a very low-calorie intake, it was one step closer to normality.

The daily routine included classes, free time, and movies (although mostly limited to comedies and Japanese propaganda). Some that could afford it bought food from vendors and built elaborate nipa shacks for privacy and separation from the masses. Our family was not in the latter group. Things took a turn for the worse with the arrival of Warrant Officer Kanishi in 1944 when the military took over the management of STIC. Rations were cut, and life started to get more difficult.

We were lucky. We had signed up to move to a new camp built at Los Banos by a contingent of former STIC men and supported by Navy nurses from Corregidor. We had volunteered to make the move to Los Banos because families could live together in nipa-constructed barracks.

Things started quite well under the command of a peg-legged Colonel. The food was adequate if one considers one pig for 2,400 people adequate. Most meals were in the form of a rice stew called *lugao*.

We were given a small plot and seeds to cultivate tomatoes and sundry other vegetables, but unfortunately were never able to harvest such because some fellow internees had gotten there before us. I did manage to harvest a handful of peanuts. Not much for our gardening efforts.

Then, history repeated itself with the arrival of Warrant Officer Konishi again sometime in 1944. Rations were cut almost immediately.

By mid-1944, things began to rapidly deteriorate as we were provided with less and less food. A couple of internees were shot as they were returning from foraging for food outside the camp. Barter for food consisted of giving whatever we had to trade to the guards, who then bought from the locals and gave us whatever they wanted. My mother's half-caret diamond engagement ring got us two pounds of rice and a couple of packs of Japanese cigarettes.

Perched precariously in a tree fifty feet from a guarded gate munching on a guava fruit too far out for the older boy to reach, I witnessed a trade effort gone awry. A guard was haggling with one of the native farmers and could not come to an amicable agreement. The guard threw down some "Mickey Mouse" (Japanese occupation currency) and turned. The farmer pulled out his bolo and took a blow to the guard's head. He chopped off one ear and ran. Shortly after that, a squad of guards double-timed out of the gate, and I am sure there was an atrocity committed in the barrio that evening.

On January 10, 1945, we were startled to find that the Japanese had left us. Food rations were increased; an American flag was raised; a radio broadcasted news and played hit songs; Bing Crosby and the Andrew Sisters belted out "Don't Fence Me In," a particularly appropriate song; and people went out of the camp to barter or buy food. George and I came back with a half sack of fruit. On my second trip, I was just slipping back under the wire fence empty-handed when the sound of boots double-timing sent shivers down my back: The Japanese were back and manning their guard posts again.

On February 22, 1945, the camp was concerned while speculating just what was the intent of the Japanese digging ditches to nowhere. We were even more distressed when we received a tuna can ration of uncooked rice and were told "There is no more food in the camp!"

The next day, it was a worried, sleepless group of internees that shuffled our way to the 7:00 a.m. roll call. The sound of the guards going through their daily exercises accompanied by their ritual grunts and groans did nothing to assuage the internees' concerns.

As we were shuffling through the dirt passageway, I told my mother I heard a strange noise. She opined it was probably a stray airplane.

As I was framed in the open entrance to the barracks, I was shocked to see these beautiful C-47 aircraft flying four to five hundred feet straight in front of me. Then some specks were seen. Someone shouted: "They are bombing!" as the specks were seen to be holding something . . . "Oh my god, it's paratroopers!"

Simultaneously, all hell broke loose. The RECON squad that had been camped close by all night burst into the camp, shooting at the guards and, more importantly, throwing phosphorus hand grenades into the room where the guards stacked their weapons. The guard posts were attacked by Filipino guerrillas; the jumpers charged into the camp. Bullets were flying all over the place.

We ran back to our cubicles and hit the floor. George, ever curious, crawled to the doorway to get a better look, only to have a twelve-inch splinter hit ten inches from his nose. He crawled back to my corner quickly. Then we became aware of the source of my imagined noise as fifty-four Amtracks rattled their way into the camp.

Meanwhile, the Army Air Corps provided air cover and flew their P-51s low over our barracks. So low in fact, I could see they had not had time to shave.

The euphoric internees wanted to celebrate and savor the moment, but LT Ringler and Major Burgess, the leaders of the flawless operation, were more interested in "getting out of Dodge" since there were two hundred Japanese Infantry just two miles away and a Division of Infantry twenty miles east and west.

He had the barracks torched, which got our immediate attention, a wonderful sight.

When the AMTRACS and the precious cargo of internees returned to Mamatid beach, behind our lines, the two hundred internees were met by a host of Filipinos bearing fruit and other foods delivered with a kind of warmth I will never forget.

Company B of the 511 Parachute Infantry Regiment, assisted by Filipino guerillas, rescued all the Los Baños internees from behind enemy lines and whisked them to safety, by crossing Laguna de Bay in the amphibious craft.

Intelligence indicated to us that there might have been plans to execute all the internees including us.

It was reported that in the rescue, the liberators killed some seventy-five Japanese soldiers and only suffered a single leg injury from the jump and a couple of minor gunshot wounds.

I think Colin Powell was quite right when he stated—almost fifty years after the raid, while Chairman of the Joint Chiefs of Staff, in a letter to the 11th Airborne Division Association—"I doubt any airborne unit in the world will ever be able to rival the Los Baños prison raid. It is the textbook airborne operation for all ages and armies."

The single most dangerous situation, at that moment, was the possibility of overeating. Supplies had to be parachuted into our temporary home in New Bilibid Prison within two days because the planned rations had all been consumed by the starving internees.

On the first day after liberation, at almost thirteen years of age, weighing in at some sixty-plus pounds, I went through twelve half rations of pork and beans, enabled partly by regurgitating some because of an attempt to savor KLIM, a rich powdered milk product.

I resurrected my penchant for visiting those wounded and sick in the hospital wards at New Bilibid. The first visits went well. One soldier had twenty-three bullet wounds in his leg but seemed quite upbeat. When I ventured to the farther wards, I first noted a strong, ugly stench, but continued through large doors and stepped into hell incarnate. The moans of severely wounded soldiers were what I heard. The smell was from the putrefying flesh of the wounded! I saw a young Filipino, about sixteen years of age, staring at me with the left side of his face blown off. I resisted vomiting but turned and ran.

God Bless our corpsmen and medical staff who see such horrors daily.

This is not a scene I will ever forget but know this is repeated time and again when war, terrorism, or domestic violence exists. We are civilized?

In April of 1945, we former internees were embarked in the USS *Admiral Eberle* (AP-123) for passage to the USA at last.

One last adventure remained. Our ship sailed through the horrible typhoon that sank three of our destroyers. A fellow internee of mine advised that one of our escorting destroyers had to turn back because of storm damage. Not surprising when I saw the Jeep carrier taking white water over the bow.

No wonder Plebe Year seemed so easy for me.

AFTER THOUGHT:

Neither I nor my family were heroes. We were survivors. The heroes were the selfless rescuers.

Forgive me please, but seventy-three years later, I still well up when thinking of the people involved in the raid on the prison and our rescue. Some probably could not legally buy beer but were already battle-tested. Their concern for the plight of each internee and the courtesy shown were wonderful to behold. It had been reported that none of that group would have sold his place in the jump for a million dollars. I believe it!

I feel the same way about the Filipinos who met us at the beach where the Alligators (AMTRACKS) landed us, with fruit, food, and love. Mabuhay!

In 1947, Dad was in a bar in Idaho Falls, Idaho, when a mustering-out paratrooper walked in. Dad asked him what outfit he belonged to.

His reply was, "511 PIR (Parachute Infantry Regiment)."

Dad asked, "What company?"

He said, "Company B."

He was the poor trooper who hurt his ankle during the jump. I believe he was one of our JONES boys.

He did not buy any drinks that night.

KGB ENCOUNTERS
Paul Roush

IN MY TIME stationed in Moscow, I sometimes traveled with a person who was not military but worked instead for another unnamed agency with headquarters near Washington, DC. His employer was no secret to our Soviet hosts, as every time I traveled with him, the KGB escorts tripled their number and aggressiveness. On one occasion, we were innocently driving along outside of then-Leningrad in the vicinity of a Soviet communication station with antennae whose type and orientation were of interest to my companion's agency.

He was driving the Zhiguli, and I had my Canon A-1 camera with the motor drive at the ready on my lap, inside the sleeve of a sweater at the proper angle to shoot past the driver and capture the scenery outside his window should something interesting materialize. A carload of "goons" was following behind us, and approached very close to us as we neared the site. Then a second car pulled out in front of us. At the precise moment my trigger finger energized the motor drive, a Soviet helicopter swooped down to a position about fifteen to twenty feet in front of our windshield and perhaps five to ten feet higher than our car and maintained that relative position as we continued to drive down the road.

One of the crew in the helicopter was filming us through our windshield with some sort of camcorder. Fortunately, I kept my hands still under the sweater on my lap and the motor drive continued to do its thing. I may have remained motionless on the outside, but I was in sheer, stark terror on the inside, and expected the "goons" were going to physically stop our car and pull us out.

I was mentally reviewing the steps for opening the Canon and pulling out and exposing its contents while at the same time, I was conjuring up images of my career going up in smoke. Instead, the helicopter pulled away and the "goons" behind us dropped back, and those in front zoomed away. The event lasted only a few seconds, but it seemed like a lifetime to me. We drove back to the American Consulate for a fresh change of underwear and continued the rest of our mission. Another day of undetected crime. Life is good. Retirement is very good.

Leningrad Seven

I was stationed from 1977–1980 in Moscow as Assistant Naval Attaché. In June 1978, seven Soviet citizens of both genders from two families rushed into the American Embassy and refused to leave. They were Pentecostal Christians from Siberia, survivors of decades of severe persecution for their beliefs, who wanted to emigrate from the USSR. Embassy staff urged them to return to Siberia but were unwilling to expel them physically. The seven remained in the Press and Culture waiting room for weeks, hoping to find a way to emigrate. My wife, Annette, who worked in the Press and Culture section, befriended them, and both of us spent a lot of time with them after working hours, bringing them food and learning about their experiences.

Finally, embassy officials removed them to a small (ten by sixteen) room in the basement and barred most folks, including us, from visiting them. Annette and I had promised we would make their story known to the outside world and would work to find a way to get them out to the West. We found an author in England who had written about a similar, though failed, attempt at the American Embassy fifteen years earlier by a larger group that included some of these same people. The author, John Pollack, came to the embassy and was permitted to interview the seven.

On his return to England, he decided to write a book chronicling their lives. Since I was not then permitted to visit them, I would pass Pollack's questions to them via the embassy chaplain, who could visit; they would write their responses for me via the chaplain, and I would mail the responses to a colleague in Paris who would translate their responses into English and mail them to Pollack.

This process continued for months with the eventual result that the book, *The Siberian Seven*, was published and widely distributed. My wife and I were eventually allowed to visit them in their room, and we spent most non-traveled days with them. We decided to try to have groups in the West take up their cause and were quite successful in doing so with groups in Scandinavia, Europe, and the United States.

When I rotated back to the States in 1980, they were still in their tiny room. We decided to continue the effort from the States and were able to enlist additional groups who would work with us. Eventually, though I was still on active duty, we decided to move the effort into the political realm. We met with members of Congress and their staffers. I contacted a then-aide to Senator John Tower of Texas, Robert "Bud" McFarlane, with whom I had served in the Marine Corps. Bud told me he would do whatever he could to help.

He later moved on to be President Reagan's National Security Advisor, from which he informed me that Alexander Haig had told Ambassador Dobrynin that the refusal on the part of Dobrynin's government to let the seven emigrate would hurt relationships between our two governments. We worked, along with others, to make the events known to a wider circle of congressmen, chief among whom was Senator Levin of Michigan.

Eventually, the Senate passed legislation granting the seven permanent residence status in the United States, effective from the date when they entered the embassy, though the State Department lobbied against it. We arranged for busloads of folks from regional churches to attend those hearings. The presence of standing-room-only crowds altered the atmosphere of the hearings.

We mass-produced slide-and-tape materials and distributed them to churches that were then able to tell the story accurately, thus enlarging the lobbying effort on behalf of the seven. We worked with a group in Switzerland to accumulate data about the Soviet non-delivery of international mail addressed to the seven in the embassy. The compiled data were taken to the International Postal Union conference by the Swiss group as a means of exposing the Soviets' violation of international agreements.

Within days, the Soviets began to deliver mail for the seven to the embassy. There is no space here to detail any but the smallest proportion of the events that occurred in the struggle but suffice it to say that on the fifth anniversary of the arrival of the seven, the Soviets permitted them and twenty-five other members of their immediate families (plus one dog) to leave the USSR.

Since 1983, they have been living in the United States. I still stay in touch with them and am pleased to say they are doing well and are very happy in their adopted country. Although there are many ways in which those five years had an impact on my life, perhaps the most important is the fact that, while I was still working at the embassy, some of the extent of my involvement became known by some of the embassy staff. The US ambassador threatened to expel me and directed my military boss to write a negative fitness report, hoping thereby to end my career.

I was nonetheless selected for colonel by the next promotion board. As a faculty member at the National War College, I led a group of students on a trip to the Soviet Union, including a stop at the American Embassy. While visiting the seven (still there after four years), I introduced myself to the wife of the Deputy Chief of Mission, who said, essentially, "No need for introductions. Everybody knows who you are. You're the hero—the one who stood up for the seven when no one else would." None of the facts of the case had changed. None of my actions were different.

Why the change in attitude? Simply the earlier events had occurred during the Carter administration with Cyrus Vance as Secretary of State, while this visit occurred during the Reagan administration with Al Haig as Secretary of State. The point for me was simply that one chooses to do what one is convinced is right without regard to which way the wind is blowing at the time. Those five years during the emigration struggle of the seven were among the most informed of my life.

LOCKED IN THE JAIL!
George Phillips

THE GOAL OF EVERY Surface Warfare Officer is to have a sea-going command. As a lieutenant, I was privileged to be assigned as Commanding Officer of the auxiliary ocean tug, USS *Penobscot* (ATA-188), homeported at the Naval Supply Center, Bayonne, New Jersey. The crew was comprised of about thirty-five officers and enlisted personnel, including my Executive Officer, LTJG "Rusty" Seacat, and two Chief Warrant Officers, Jim Fagan Chief Engineer, and Boatswain (Boson) Bean, First Lieutenant.

I could not have handpicked a more competent wardroom. The remaining crew members were comprised of highly qualified petty officers who had been carefully screened for independent duty.

As an aside, Rusty was from Kansas and had never seen the ocean before joining *Penobscot*. Jim Fagan had been the senior "Motor Mac" of the PT squadron stationed in the Philippines during the outbreak of WWII. It was his squadron that is credited with taking General MacArthur and his family out of the Philippines to a waiting submarine, and hence to Australia. I understand the General and family were violently seasick on the PT boat journey. Boson Bean was the antithesis of his rating, with no tattoos and the most profane words spoken were "gizzard head" and "dangnabit."

The single most memorable mission in the two years I had *Penobscot* was a tow of a mothballed destroyer from Philadelphia to Orange, Texas. Rounding Cape Hatteras was somewhat exciting in that we took a fifty-degree roll, a very scary situation. Our damage control data book stated that fifty-two degrees was the point of no return. Immediately thereafter, we learned of a hurricane forming off the Florida coast.

I decided to evade the storm and seek shelter in Port Everglades—now called Fort Lauderdale. About a day out, I began to experience pain in my lower abdomen which intensified as we neared the port. After successfully mooring, I called the corpsman, explained my symptoms, and after a cursory exam he sent for an ambulance, and off to the hospital I went. It was not appendicitis as the corpsman had thought, but rather gastroenteritis. The doctor informed me that this was as close to experiencing childbirth as a man can get. With treatment, the symptoms abated, and I was discharged the following day.

Upon returning to the ship, I was updated on the hurricane's intensification and predicted path, thus mandating we should remain in port for several more days. The crew had not been paid for several weeks and were making it known that liberty in Port Everglades was rather tame without any money. My inquiries determined that if the commanding officer were to take the crew's pay records to Coast Guard, Miami, they would make timely disbursements.

The next thing I was made aware of was that a crew member had managed to check out a Navy pickup truck that was presently sitting on the pier for my use. I soon found myself on A1A, a four-lane highway with storefronts on the right-hand side. A local driver had stopped at one of those stores and apparently, without looking, backed out into my lane. Instinctively to prevent T-boning him, I swerved left, which caused my front bumper to leave a long scratch on the car passing me in the outside lane. We both stopped. He was understandingly irate, called the police, and I was given a ticket. To ensure I would be in Miami traffic court that afternoon, the officer took my driver's license.

So, on to traffic court, I went. Upon entering the courtroom, it was soon apparent the judge was bilingual, as he handled the vast number of defendants in Spanish. When it came to be my turn, the charges were read out and the judge asked, "Guilty or not guilty?"

I responded, "Not guilty."

To which the judge replied, "Twenty-five dollars."

I again stated, "Not guilty."

The judge responded, "Trial to be scheduled . . ."

But I interrupted him by stating, "I'm in the Navy and can not guarantee—

Before I finished my sentence, the judge ordered, "Bailiff take him away," and away I went, escorted out of the courtroom, down a hallway, and into a small room.

I heard the door lock behind me, and I was furious. The lone furnishings were a pay telephone on the wall and a single chair nearby. After a few minutes, I cooled down and realized I had some loose change in my pocket. I used the lone nickel to call the ship and asked the XO to come to the courthouse with twenty-five dollars.

He replied that he did not have that amount of money, nor did he think anyone in the crew did.

I told him to pass the hat—certainly, there was twenty-five dollars among the crew.

An hour or so later the door was opened. Standing in the doorway was the Bailiff, XO, and CWO Boson Bean in his dress blues as if ready for Captain's Inspection. When I asked what he was doing there, he replied, "I never had the opportunity to bail out the Old Man, and I wasn't going to pass this one up." Funny!

As we were walking to the clerk's office to pay the fine, we came abreast of the courtroom just as the doors opened. There stood the judge. He and Boson made eye contact and literally froze in place for at least ten seconds. They then ran toward one another, embraced, and jumped up and down like two teenagers at a rock concert. I just stood there, open-mouthed, having no earthly idea of what to make of the scene playing out in front of me.

After a minute or so, the two separated and Boson told me he and the judge were from the same hometown, went to high school together, and after graduation enlisted in the Navy, went to recruit training, and were ordered to the same battleship stationed in Pearl Harbor. They were manning a forty-millimeter anti-aircraft gun on December 7, 1941, when an explosion from a Japanese bomb blew them over the side and into the water. Each assumed the other had perished. Now, twenty-four years later, they were reunited. WOW. The judge then turned to me and asked why I was there. After a short narrative, he said, "Court costs, five dollars."

That afternoon from the peace and quiet of my stateroom, I called my wife. When she answered the phone, I asked her to sit down. I then said, "Well, dear, in the past two days I have been in and out of the hospital and in and out of jail." What else was there to say?

DEPENDENT'S CRUISE
Eleanor Boyne, Wife of Peter Boyne

THESE INITIAL PARAGRAPHS are a preamble to set the scene for the particular adventure described here.

Peter reported to his first submarine in late 1958. The USS *Cavalla* (SSK-244), home-ported in New London, Connecticut, was a Gato-class submarine, best known for sinking the Japanese aircraft carrier, *Shokaku*, on her first patrol during WWII. She went on to make six patrols before the end of the war, and she was in Tokyo Bay for the

surrender ceremony. In 1960, there were a number of the crew aboard who had made WWII patrols and wore the war patrol pin along with their dolphins.

Peter's first Skipper, Cdr. Lowell "Slick" Fitch, on *Cavalla*, wore a patrol pin as did SD1 Ragland who served with Cdr. Eugene "Lucky" Fluckey on *Barb*, the boat that "sank" a train. It was not uncommon for senior officers who served during WWII to have earned nicknames for various exploits. For example, Adm. Charles Randall "Cat" Brown, father of a classmate. Or, Adm. Frederick "Fearless Freddy" Warder, who was ComSubLant. The crew were never called by their first names, but I fondly remember SD1 Ragland, EN2 Sullivan, EN3 Cockerton, EN1 McAllister, and HMC Scott, Chief of the Boat. *Cavalla* was placed out of commission in 1952 for conversion to a hunter-killer submarine and reclassified as (SSK-244) (K for *killer*!). The conversion included remodeling the bow with the addition of a curved housing for the BQR-4 sonar system. The original conning tower and bridge were converted to the classic sail configuration seen today. Her new sonar made *Cavalla* valuable for experimentation, and she

was transferred to Submarine Development Group 2, New London, to evaluate new weapons and equipment, and to participate in fleet exercises. There was a book published around this time entitled: *Where Did You Go? Out? What Did You Do? Nothing.* This clearly describes *Cavalla's* classified at-sea periods. She was decommissioned in 1969, and in 1971, *Cavalla* was transferred to the "Texas Navy" as a museum ship in Seawolf Park, Galveston.

It was the custom to organize a Dependent's cruise for families—mostly for the wives—so they could experience a day at sea and observe the boat's crew in action. On a bright, sunny day in late summer of 1960, *Cavalla* was put out to sea, accompanied by USS *Angler* (SS-240). The boats sailed down the Thames River, under the New London railroad bridge, out into Long Island Sound. It was the procedure for the bridge to be raised to allow the boat to sail through. However, if a train was due to cross, the boat was required to tread water until the train passed. The Skipper at the time, Lcdr. Robert "Yogi" Kaufman, had no patience to wait, especially when returning to Base. He wanted to get home; so, he would order the boat to "flood down" and sail under the bridge. The crew became very competent in the procedure, and *Cavalla* became famous for this maneuver when returning to port.

The boat's guests were treated to coffee and snacks in the wardroom and provided a tour of the boat—the torpedo rooms, the engineering spaces, the galley, the maneuvering room, the "head," and berthing spaces. The young sailors proudly manned the dive planes, the ship rigged for dive, called a "green board" and submerged. While at periscope depth, *Cavalla* fired a water slug from its torpedo tube. One quickly got the idea of what it was like to fire the real thing as the kick-back from the firing was impressive. Surfacing was exciting as the diving officer put a sharp up-angle on the boat and the guests braced themselves to keep from falling over. On the surface once more, the safety lines were rigged and the guests were allowed topside. *Angler* had been shadowing *Cavalla* and now it was time for another demonstration—a real torpedo shoot. What? Of course, *Angler* set her torpedo (unarmed) to run under *Cavalla*, but seeing that golden tube speeding toward the boat made one a bit apprehensive. A sight I will never forget.

Several months prior to the dependent's cruise, the boat's Executive Officer, Will James, died unexpectedly of a cerebral hemorrhage. The Catholic Chaplin on the Base was Lt. John (Jake) Laboon, and he accompanied the group for the day at sea. Father Laboon graduated in 1943 from the Naval Academy and served in the submarine force during WWII. He resigned his commission after the war and entered the Society of Jesus (Jesuits). Upon ordination, he applied for a commission in the Reserves, and shortly thereafter was recalled to active duty. He celebrated Mass in the forward torpedo room in memory of Will. A very special experience for those who attended the Mass. The USS *Laboon* (DDG-58) is named in his honor.

Time aboard *Cavalla* earned the guests the title "Honorary Submariner" and all were awarded qual cards signed by the Skipper. Everyone agreed that it was a day well spent.

FRITZ, A MARINE IN VIETNAM
Fritz Warren

IN THE SUMMER OF 1964, I completed a course in Communications/Electronics at the Naval Post Graduate School in Monterey, California. Afterward, I was posted to the Marine Corps Development Center at Quantico, Virginia, for a three-year tour and was assigned to the Communications/Electronics Division. When Marines landed in Vietnam in May of 1965, I immediately became impatient with the remaining two-plus years left on my tour. I submitted a request to be released to serve immediately in Vietnam. Marines were in a combat situation, and I wanted to be there also! I got as far as General Hurst, the Commanding General of the Development Center, and he, while sympathetic, informed me there was no one else in the pipeline to replace me in the important technical job for which I was responsible. Shortly after I failed to escape, I was also made the Marine Corps Project Officer for tactical secure voice systems, served the remainder of my tour, and received orders to the Ninth Marine Amphibious Brigade (9th MAB) in Okinawa.

Before my arrival, I sent the customary letter to the Commanding General of the 9th MAB indicating the date of my arrival and my willingness to serve in any capacity where he felt I could be most useful to his organization. In Okinawa, I reported to the G-1 (Personnel), as the CG was "off-island." The G-1 was somewhat pudgy in appearance and less than a picture of a Marine. He told me I had been assigned as the CO of Headquarters and Service Company, an organization he said was essentially responsible for taking care of the disciplinary problems of Marines who were unable to stand up under the pressure of combat and who were awaiting various types of action, ranging from courts-martial to administrative discharge from the Corps. I was somewhat shaken by the news of my assignment, and then I recalled how Captain William Weise, my company commander in 1958, handled a similar situation and decided to follow suit.

I told the LtCol G-1, "I could outrun, outfight, or out anything anyone available for a job in-country with one of the two Special Landing Forces that were embarked aboard amphibious ships operating off the coast of Vietnam."

He stood up and seemed quite shocked, but managed to say, "Major, the line for those jobs is very long."

The line had already been cast at this point, so I retorted, "Bring me to the head of the line, and I will knock the first guy flat on his a-s."

At this point, the slightly rotund G-1 ordered me to follow him to the office of the Chief of Staff. When we arrived, I was instructed to sit down while the G-1 spoke with the Chief of Staff. I was a little worried at this point about what was about to transpire and could envision myself "in-hack" for a fortnight because of my insubordination. After about fifteen minutes, I was ordered into the office of the Chief of Staff. I was not offered a chair, so I stood at rigid attention. The Colonel told me, "Major Warren, I want you to keep your mouth shut and not say one word." He then said he was shocked at the words I used when talking to the

G-1. Then, to my surprise, he told me I was being assigned as the Assistant Operations Officer for Special Landing Force Alpha (SLF A) which was embarked on Amphibious Ready Group Alpha (ARG A) ships off the coast of Vietnam. I could hardly contain myself but somehow did. I have always been grateful the Chief of Staff was a naval aviator and could appreciate the message in my outburst, whereas if he had been an infantry officer, I feel sure I would have never left the island of Okinawa for the duration of my tour.

The next day, I was off to Danang to grab a ride to the USS *Princeton*, the flagship of ARG A and headquarters of the SLF. I worked on the detailed planning for five landings while a member of the SLF staff before being reunited with LtCol William Weise, who was now the Commanding Officer of BLT 2/4—which is shorthand for the Second Battalion, Fourth Marine Regiment.

My CO was a wonderful leader—LtCol Bill Weise. I had earlier served with him when he was a captain, and I was a first lieutenant. As the operations officer for BLT 2/4, I was now responsible for coordinating the other staff members and for developing plans for the combat operations against the North Vietnamese forces. For the next six months, I was located physically near the Demilitarized Zone and most of the time north of the Cua Viet River.

The BLT consisted of a Marine battalion (2/4) and several attached combat support organizations, making it capable of limited independent combat operations. During the months of January through April of 1968, BLT 2/4 was heavily engaged in combat action against enemy soldiers of the North Vietnamese 320th Division. We suffered many killed and wounded Marines because of the constant fighting that took place while 2/4 defended the northern bank of the Cua Viet River. This river was important to the Marines, and it was used to transport more than 90 percent of the ammunition and supplies used by the US forces serving in the area. BLT 2/4 was supported by aircraft, artillery, and naval gunfire from ships in the South China Sea. In this period, we experienced continued battles with the North Vietnamese soldiers who were trying to move through our area to get into positions where they could cut off the flow of materials on the Cua Viet River and attack the Marine base at Dong Ha.

On April 30th, one of the US Navy boats on the Cua Viet River was fired upon from the north bank of the river near the village of Dai Do. For the next three days, BLT 2/4 moved its Marines into positions where they could push the enemy soldiers away from the river so the Navy boats could resume their operations and movement of supplies. The battle was hard fought by both the Marines and the North Vietnamese soldiers, and it lasted for three difficult days. The Marines were under-strength from their constant contact with the enemy over the past three months (normally a rifle company has 206 men, whereas, at the beginning of this battle, the average strength of each rifle company was approximately 165 Marines.) At the end of the three days of combat, 2/4 had lost about eighty-five Marines to enemy fire and another five hundred had been evacuated due to their serious wounds. Each company had received some reinforcement from Marines on board the ARG ships, but even with these, each of the four companies had an average strength of only about twenty-five men who were exhausted from their three days of constant battle.

The Battalion CO, Bill Weise, had been wounded on May 2nd, as was the CO of E Company, Captain Jim Livingston. Another company CO had been seriously wounded on the first day of the battle and evacuated to a hospital ship, and a third company CO had been wounded five times by the end of the third day. We used support from aircraft, artillery, and naval gunfire to help Marines battle the much larger 320th Division of North Vietnamese soldiers. The North Vietnamese force had excellent artillery support from units operating inside the DMZ. Keith Nolan describes the story of this battle in the book, *The Magnificent Bastards*. LtCol Bill Weise also describes the battle in the video, "Memories of Dai Do."

For the first two days of this battle, I was in a support role, constantly on the radio with higher headquarters arranging for artillery, close air, and naval gunfire support. It was a frustrating job; I was aware of the serious battle taking place because I saw the killed and wounded Marines being evacuated. On the second day, I moved with my "bravo" command group from our normal base to a position on the edge of the village of Dai Do. I had been monitoring the radio and realized the battle was fierce and that the more numerous enemies was counter-attacking against our Marines. Shortly after arriving, I saw Captain Livingston being carried from the field, he had been seriously wounded but refused to leave his men until he was unable to continue. Shortly thereafter, I learned that our CO, Bill Weise, had been seriously wounded and was being evacuated from the battle area.

At this time, as the senior Marine, I assumed responsibility as the BLT CO. I called for supporting arms to hit the enemy positions while organizing the roughly one hundred or so Marines that had survived the battle thus far into defensive positions. The enemy forces, also badly depleted, decided to withdraw to reorganize their men. We took the opportunity to collect our dead and wounded from the battlefield and to get the wounded to medical treatment. We received small attacks from the enemy forces throughout the night, each of which was successfully repelled. On the morning of May 3rd, BLT 2/4 was relieved by the Third Battalion, First Marine Regiment, and returned to its base camp.

For this action, Captains Vargas (F Company) and Livingston (E Company) received the nation's highest award, the Medal of Honor. LtCol Weise received the Navy Cross (second highest combat award) along with SgtMaj Malnar (SgtMaj Malnar was killed while protecting Bill Weise during an enemy counterattack). The Marines who fought at Dai Do earned numerous other awards. I was awarded a Legion of Merit with combat V.

I stayed with 2/4 for another three months, as we reorganized and fought other battles near the famous Khe Sanh Combat Base, which is close to the border with Laos.

In retrospect, we are left to speculate about the true nature of the Vietnam War and its effect on the United States and the countries of Southeast Asia. For my part, I did my duty as I understood it, and felt I was assisting people who wanted to govern themselves using a democratic process rather than being dominated by the Communist government from the North.

GET OFF THAT SUBMARINE
Wilson Whitmire

THE DAY AFTER RELIEVING to be Commanding Officer of the submarine *Volador* in San Diego, we set sail in company with the submarine *Tiru* (Tommy Warburton, YG'56, commanding), for the new home port of Charleston, South Carolina. We had a great port visit to Acapulco before transiting to Rodman, Panama Canal Zone (PCZ) for a three-day stopover before passing through the canal.

Rodman has tides averaging about fourteen feet, so line handlers on a vessel tied to a pier are required to adjust mooring lines around the clock, but anyone moored to another vessel doesn't have that problem. As Tommy Warburton was senior, *Tiru* entered port first and moored alongside a pier, and I intended to tie up to *Tiru*, thereby avoiding the line handling hassle. The Panama Canal Zone pilot wanted *Volador* to moor to a separate pier as he didn't think there was room between *Tiru* and a ship moored to the opposite pier.

We had an obvious disagreement, and after I finally told the pilot that *Volador* was going alongside *Tiru*, he reported the same to his headquarters, and his supervisor yelled over the radio, "Get off that submarine, get off that submarine immediately." (The PCZ and Suez are the only waters where pilots, i.e., the canal companies, are responsible for damage to naval ships—in all others, the CO is responsible).

The pilot vessel came alongside, and the pilot departed, thereby absolving the Panama Canal company of any subsequent damage to or caused by the *Volador*. We proceeded to moor alongside *Tiru* without incident and thought that was the end of the episode until, while still doubling the mooring lines, a black sedan came rolling down the pier, and I was summoned to the headquarters of the Commander Fifteenth Naval District.

Once there, I received a "talking to" from the Admiral who ended with something to the effect, "Well, Whitmire, how do you feel now about being the only naval ship on record to moor in the PCZ without a pilot?" I allowed that, right then, I wasn't feeling all that great about it. But being a submariner himself, I think deep down he understood because when I departed his office, he gave me a handshake and a wink.

The first night in Rodman, Tommy Warburton and I were joined at the Officer's Club by Charlie Flather (Brown Univ YG'56) who was CO of a submarine headed for San Diego. After a few beers, Charlie asked if he could borrow one of our anchors. We at first thought he must be joking but he was serious. His sub had lost its anchor in Lake Gatun, between the series of locks, and he was worried about a scheduled port visit to Acapulco where his submarine would be required to anchor.

Neither Tommy nor I was about to "lend" an anchor to anyone! After scouring the PCZ for an anchor, Charlie's crew, in desperation, finally bundled a few steel railroad car wheels together and tied them to the anchor chain. Later, when Charlie and I both had duty in the Washington area, he admitted that until his Acapulco visit, he didn't understand the purpose of an anchor's flukes (think maybe he slept through some OCS classes?). It seems his submarine dragged those railroad wheels all around Acapulco's harbor!

The rest of the trip to Charleston was uneventful except for a misunderstanding in Kingston, Jamaica, when the liaison to the local consulate, a marine captain, informed us that the tax-free quantity of booze was one case per person. After loading the non-returnable beverages aboard, I was informed by the marine liaison that he had erred, and the limit for a three-day port visit was only one gallon/person. Not wanting to throw or give away the above-limit bottles, I had them stowed in weighted bags in torpedo tubes—the idea being that if customs in Charleston had been alerted to our cache, we would jettison it alongside the pier. My new Division Commander, who had a vested interest in the cargo, knew of my plans but just told me to do whatever I felt comfortable with. As it turned out, we went through customs without a hitch, and I've always considered the name of the scotch whiskey I purchased to be apropos: "Old Smuggler."

GUNSHIP HISTORY IN LAOS
Brad Parkinson

THE DARK NIGHT sky was laced with narrow white lines. I was looking out the five-foot square hole in the side of the AC-130 gunship and viewing, with some dismay, those bright tracer paths streaking by us from the North Vietnamese (NVN) fifty-seven-millimeter anti-aircraft guns (AAA). The enemy, 10,500 feet below us, was trying to shoot our gunship out of the sky.

With our forty-millimeter cannons, we were methodically destroying trucks along a narrow jungle road. Suspiciously, the vehicles were stopped, with their engines running. Bait? It looked very much like a prepared trap by the NVN, but we were not backing off—they were equipment supplying vehicles for the enemy's war effort. Our mission Commander, and veteran of hundreds of missions, was Major Ron Terry. We would not leave until all trucks had been destroyed.

It was my first combat mission, and it was not clear this situation was totally normal, although I had expected something similar. Technically, I was still assigned as a Lieutenant Colonel and Deputy Head of the Department of Astronautics and Computer Science at the US Air Force Academy. *How was it then, that I was flying combat, eight thousand miles away from USAFA, over the Ho Chi Minh trail in December 1969?*

It all began three months earlier, when a friend and USNA 1957 classmate, Major Rick Willes, burst into my office at the Air Force Academy. He explained that there was an opportunity for me to help the Vietnam War effort. He had his PhD from MIT, was a fighter pilot, and was a professor in the aeronautics department, which was located near my office on the top floor of Fairchild Hall. He explained that a new version of the airborne, interdiction gunship was being developed with a first-of-a-kind digital fire-control system. The fire control was to

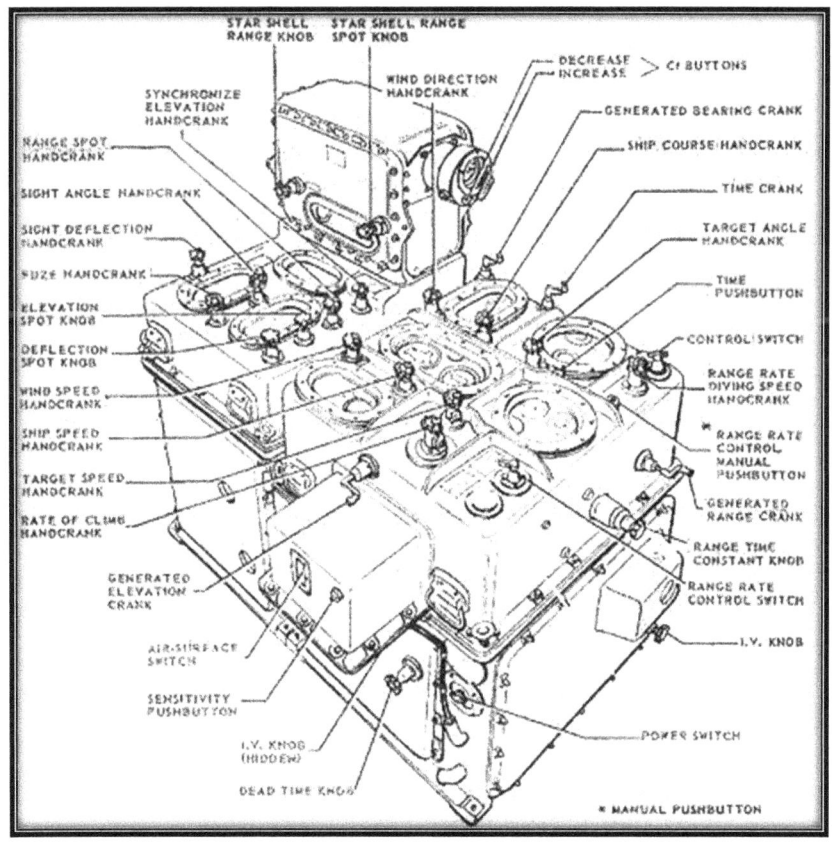

MK! Gun Fire Control System (about 1956).

precisely aim forty-millimeter cannons sideways, as the gunship circled above North Vietnamese nighttime supply traffic in the jungles of Laos. The main targets were trucks with ammunition, oil, and supplies flowing through the Ho Chi Minh trail to the western side of South Vietnam.

The gunship developers had committed to a brand-new, first-of-a-kind digital fire-control system. This was the system that calculated how to point the aircraft so the guns would hit the designated target. *It was not working*. My background included fire-control classes at the Naval Academy (for example the Mark 1 analog fire-control computer for sixteen-inch guns we had studied first class year, senior year), a formal fire control class at MIT, and Kalman estimation theory. Rick flattered me by saying I was the best person in the Air Force to get them on track. If they failed, the whole development effort would be canceled.

I was explaining to Rick I could not do that, I had just arrived at USAFA and begun teaching a course in space mechanics, and I also had other department responsibilities when he interrupted me. "Don't worry about that, I already went to the Dean, General Woodyard, and he said he thought it was a great idea." The fact he had gone two levels over my head and not warned me made me a bit peeved, but his usual enthusiasm continued unabated. After a protracted explanation, I said I would think it over.

One thing led to another, and I soon ended up at Eglin AFB (FL), working fourteen-hour days that alternated between a.m. flights in our AC-130, followed by creating digital programming fixes from about noon to the wee-small hours. After many weeks, the fire control still wasn't working properly, and, ominously, a team of Pentagon officers, headed by a General, were coming the next day to inspect the results. They knew of the failures and were evidently seeking an excuse to scrap the whole thing, ordering conversion back to the old, less accurate analog system in the already deployed models of AC-130.

I worked through the night. At about dawn, we loaded the last attempt to fix the problem into computer memory. I was supposed to fly with the mission, as usual, but I could barely walk. The crew loaded up, and I said I would see them on their return and hit the Officer's Quarters to sleep. After two  hours of restlessness, I was back on the flight line watching them land. The test had been to fly out to a small raft in the Gulf of Mexico and shoot at it from about a mile and a half away as the AC-130 circled at about 260 knots speed.

The body language of my crewmates as they bounded out of the parked plane told the whole story. "We shot the hell out of the raft! The mission was a total success!" Of course, credit went to all  of us, including the three technicians and engineers who helped me with the fixes. The General decided on the spot: deploy it immediately. Button her up and fly it to return to the Sixteenth Special Operations Squadron in Ubon, Thailand. It would join the seven older versions of AC-130 gunships that had been left in Thailand after the last dry season. But the new "hunting season" was on.

The kicker was that we were not certain all the bugs were fixed. To ensure success, it was decided our help was still required. Rick and I would fly out to Southeast Asia about a week later and join up with our plane. Combat would begin immediately, and we would augment the thirteen-man crew for the five-hour mission. My essential role would be to fly as the Fire Control Officer. Later, after I had mission experience, I twice flew as the Mission Commander.

Our new AC-130, called *Surprise Package*, had been modified with thirteen new systems. The total modification took about five months. It turned out that this single airplane was the most effective in the history of the war to that point.

Missions were a dangerous duel between the ground NVN AAA and our gunship as we targeted the truck traffic our sophisticated sensors had found. Flying at night gave us an advantage of stealth, but it also made the enemy traffic more difficult to spot. Since they only operated for about seven hours after dark began, we were constrained to use that as our hunting period. Most dangerous for us were the peaks of the twenty-eight-day lunar cycle, when a near-full moon would illuminate our plane and help the accuracy of the ground gunnery. Our policy was to press the attack, despite the enemy's increased AAA accuracy.

We almost always found plenty of targets. Typically, our mission would end when our ammunition was exhausted. We would return, tired and tense, at about two a.m. The routine was to immediately go to the officer's club for a couple of "Tooney-makers" (Heinekens and a Tanqueray Martini) and a pizza. We would then drive to our small hotel in town since there were no available quarters for our temporary-duty team on the base. By noon, we would be back on the

flight line, trying to remedy any deficiencies I had seen in the previous night's combat experience. Then at dusk, we would take off to the Northeast for another night's combat. As darkness fell, we would be back over the many faint trails, using our sensors to locate the enemy resupply trucks. There were no weekends. At one point, we had flown seventeen nights without a break. Only when an engine malfunctioned did we get a single night's reprieve.

Having quarters off the base led to one exciting episode. Normally the small town (central city population less than one hundred thousand) of Ubon, as we drove through it in the wee-small hours, would still be quite lively with people in the streets having a good time. One night, the streets were unusually and deathly quiet—no one was about. It struck me as strange, but I collapsed into bed and instantly was sound asleep. About an hour later, the phone rang, and I groggily answered. The caller informed me the base was under direct ground attack, with infiltrators who had moved across the nearby border with Laos. We were ordered back to base, although it was not clear what our role was to be. We jumped into the small pickup truck we were using, and the two of us riding in the back bed of the truck were alert in case someone tried to drop a satchel charge in with us.

No such charges appeared, but, when we arrived, the base was frenetic. Flare ships had been illuminating the whole base, particularly the far side, closer to the Lao border. The intruders did have satchel charges, and they were specifically targeting our squadron of gunships, which included our *Surprise Package* and the seven older models. We could hear sporadic, but sometimes intense gunfire. Of course, the defense of the Royal Thai Airbase was in the host's hands at that point.

In a lull between the bright flares, our primary pilot and I crept out toward the runway to see the action. The shooting was continuing on the other side. Crouching down, we felt we were virtually invisible. Suddenly a replacement parachute flare lit the whole base like daylight. Six-inch grass did not conceal our locations in the middle of the open field. On hands and knees, we slithered to safety. Curiosity could have killed these cats—dumb cats! Activity continued for several hours and, sometime around dawn, the gunfire had ceased, and we were told the base had been secured. All the attackers were dead or captured.

It seemed that somehow the Ubon citizenry had a premonition about the attack, which probably accounted for the deserted streets we had encountered earlier.

Interrogation confirmed the targets were specifically our airplanes, not the F-4s that were present in large numbers. For the NVN to mount such a desperate and expensive (for them) operation was flattering in a way. Clearly, it was a recognition (tribute?) to our squadron's effectiveness. It also brought the ground elements of the war unpleasantly closer.

Surprise Package averaged over 7.3 trucks destroyed per mission. For example, in 112 sorties, this single aircraft had destroyed or damaged 822 trucks, while the F-4s, with 6,310 sorties, had 1,576 trucks. In the effectiveness per sortie, the ratio was about thirty to one.

I ended up with twenty-six missions and over 150 hours of combat time. Our plane was about twice as effective as the previous models, which were still operational there. Our crew set a record (at the time), of thirty trucks destroyed in a single mission. There were numerous personal and unit awards doled out, including the Presidential Unit Citation to our Squadron, the Sixteenth Special Operations Squadron.

Later (1972), in researching a paper at the Naval War College, I singled out the records for my first night of combat. I discovered it was the worst night in the whole year's campaign for NVN fifty-seven-millimeter shells, targeting a single gunship. It was much more normal to be attacked by the less effective thirty-seven-millimeter guns. Not proven, but it certainly reinforced the suspicion that we had entered a trap that night—fortunately, emerging unscathed. In fact, we had violated our instructions and fired "Counter-Battery" until the fifty-seven-millimeter AAA guns had been silenced. In subsequent years, the NVN increased the number of fifty-seven-millimeter installations and downed some of our gunships.

> Surprise Package shattered all 16th Special Operations Squadron records on February 14, 1970, by destroying forty-three trucks and damaging two in a single mission. Successes like this enabled the unit to claim its 5,000th truck destroyed or damaged on February 21, 1970.[29]

Shortly after my return and resumption of my position as Deputy Department Head, I received good news. The *Surprise Package* plane, with a regular crew, had greatly surpassed our earlier thirty-truck record:

Ironically, the forty-millimeter gun model we were using was identical to the anti-aircraft "Bofors" gun I had loaded during firing drills, on my first Midshipman cruise on the battleship *Missouri* in 1954. In fact, the ammunition we were using, on the AC-130, was stamped 1944, and had been manufactured during WWII, twenty-five years earlier! While we had a few "duds," most of the rounds were just fine. By the next season, a 105-millimeter field howitzer (4.1-inch diameter) was added to the armament. Others in my department participated in the evolution, but I had moved on.

As the rainy season began, the truck traffic would cease which had been the trigger for me to return to the Academy. For the following, 1970–71, year, I was appointed as Head of the Department of Astronautics and Computer Science and had no more combat adventures. But as the Ho-Chi-Minh traffic resumed in that year's dry season, I would shudder when the monthly, bright, full moon illuminated the Colorado night skies. I would recall flying the trail, with such a full moon silhouetting our plane against the dark-sky background. The NVN gunners would then have a very clearly defined target and an advantage in the deathly duel that was refought each night. That full moon would induce those powerful memories; I could not help reflecting on the isolated planes with their thirteen-man crews at extreme risk. From 1969 to 1972, seven gunships were lost, with over fifty crewmembers. Effectiveness came at a cost.

Our *Surprise Package* AC-130 began a love-hate relationship with the US Air Force. Initially retired to the reserves after the war, large gunships have been repeatedly brought back and upgraded with brand-new airframes and improved equipment. The latest model is the AC-130J, deployed with the Seventy-Third Special Operations Squadron at Hurlburt Field, Florida. The Air Force plans to have thirty-seven active aircraft by FY2025.

It is gratifying to recall I had a small, cameo role in early AC-130 history and acceptance. These aircraft are truly a precision weapons delivery system. More than anything else, that precision is due to the digital fire control system with its phenomenal accuracy.

HIGHLINE ADVENTURE
Ron Goldstone

D ESTROYER SQUADRON EIGHTEEN deployed to the Mediterranean with a full complement of officers: the Chief Staff Officer (that was me), a Communications Officer, an Engineer Officer, his staff Doctor, and a Chaplain.

The Commodore (our boss) was a personable type, eager to observe his Commanding Officers, up close and personal, to evaluate their effectiveness. He took advantage of every opportunity to get onboard their ship. Whenever one was in company with the Flagship, the Squad Dog (the Commodore) would arrange to pay a visit via Highline. This is a line between two ships where someone is moved from one ship to the other in the Highline chair. Normally, he would go alone so as not to create a burden on the target officer's wardroom.

On one occasion he broke from his usual pattern and invited the Chaplain and the Doctor to accompany him, probably to give them the experience of riding the chair across.

As I was observing the transfer from the bridge, a signalman sidled up and asked what was going on. I told him the Commodore was paying a visit to the other ship. When he noticed that several others were also making the trip, he asked who else was going over. I explained that the Chaplain and the Doctor were with him.

After a pregnant pause, the signalman, wryly opined: "Seems to me like the Commodore is a God-fearing hypochondriac."

LEFT ON THE BRIDGE
Jim Beatty

SOMETIME DURING the early to mid-1960s, USS *Halfbeak* (SS-352) was at sea conducting diving operations. On one practice dive, a dog (the part of the hatch that holds it down) on the upper conning tower hatch broke as the Officer of the Deck slammed the hatch down as he left the bridge. To avert disaster, the newly qualified Quartermaster of the Watch joined the Officer of the Deck on the ladder to the bridge in holding the hatch down and keeping it shut until we could surface. The QMOW and the OOD got wet but the rest of us, except maybe the helmsman directly under the hatch, stayed dry. As I recall, the submarine tender maintenance crew on the Fulton fixed the hatch, the QMOW got a well-deserved Bravo Zula (well done), and shortly thereafter, *Halfbeak* headed to Bermuda. During the port call, two events that contributed to this yarn happened. A few ROTC Midshipmen came aboard for a training cruise, and I bought a pair of sea boots.

A few days later, we were at sea conducting dives. I was the

Officer of the Deck, there were two dungaree-clad lookouts, and there was a tall, gangly, khaki-clad, wide-eyed Midshipman on the bridge with us. I explained to him the process of clearing the bridge before diving the boat. I had also decided he would probably be a little slow on the ladder getting below. The Quartermaster, the recent hero, was busy at that moment at the far end of the Conning Tower plotting a fix. The Skipper, then Lieutenant-Commander Frank Adams, stuck his head up the hatch and gave me the order to "take her down." I increased speed, gave the order to clear the bridge, and while watching the quick lookouts and slow Midshipman drop through the hatch, sounded the diving alarm two times.

To my surprise, the hatch slammed shut after the Midshipman went down the ladder. I rang the surfacing alarm and shouted over the main communications, "You left me on the bridge, you left me on the bridge," and stepped back to see if I had beaten the Interior Communications Electrician to the cutoff switch. I remember thinking about crawling up the cowling and I didn't want to get my new seaboots wet!

A few observations and comments. The noise of the air rushing out the ballast tank vents was loud but fortunately for me, brief in this instance. It was very quiet on the bridge after the diesel engines shut down, and we went on battery power. *Halfbeak* didn't settle very far down into the ocean before the welcome sound of high-pressure air to the ballast tanks and the rumble of the low-pressure blower was heard to get the water out of the ballast tanks.

A brief time later, the upper hatch wheel turned, and the hatch opened slowly. Captain Adams appeared with a serious look on his face. I remember saying, "Good drill, Captain."

He quietly responded, "That was no drill." It seems that the QM was not aware the Captain had ordered a dive and didn't realize a Midshipman was on the bridge. He raced from the far end of the conning tower when he heard the diving alarm, saw two sets of dungarees and a set of khaki go by, and observed a wide-open hatch. He did the right thing by slamming and dogging the hatch. (At least I don't think he had it in for me!)

All is well that ends well. Besides, I have a great sea story that has been told many times.

MESSAGE IN A BOTTLE
Earl Piper

THE TIME: AUGUST 1949. The occasion: departing New York Harbor. My family was aboard the SS *America* sailing to Southampton, England. My father, a Marine Officer (USNA 1927), was under orders to report to the US Embassy in London. My younger brother and I loved to explore the ship. About a day out of New York, we found an empty green ginger ale bottle. It gave us an idea. *Let's put a note in the bottle, cork it, and toss it off the stern into the wake.* This we did. On the note, I put my name and the Fleet Post Office address where we would live for the next two years.

We had beautiful weather all the way across the Atlantic during the six-day crossing. When we arrived in England, our family settled into a row house located in West Kensington, London. My brother and I attended British boys' schools while we were there. The city still showed many signs of the "Blitz" of WWII. There were still large areas of rubble and half-destroyed buildings, and some food was still being rationed.

My father's tour of duty approached its finish in the early summer of 1951, and we looked forward to returning to the States. I was fifteen. One afternoon, my father came home with a letter in his hand. It was a blue envelope with French stamps and addressed to me. My father asked, "Whom do you know in France?" I told him I knew nobody there. We excitedly opened the letter. It was written in French. My schoolboy French was good enough to read that a little girl and her father had been walking along the beach close to their home in Landes, France, on the Bay of Biscay and had found a green ginger ale bottle that had washed ashore with my note inside.

The girl was four years old. Her name was Marie Christine Deleste. The letter to me had been written by her schoolteacher and signed by each of the eight students in her small class. They wanted to know who I was, my age, and where and when I had thrown the bottle into the ocean. They asked for my picture and seemed thrilled to be able to share news of life in rural France and to correspond with this mystery person from afar.

Several letters, two or three each, were exchanged during that early summer of 1951. I still have those letters, a picture of Marie with her daddy and their little dog, and the actual note that went for the long ride in that bottle; they provided the actual note in their second letter.

A bit of recent homework reveals that the bottle drifted over two thousand miles in a period of some twenty months while being pushed along by the Atlantic Gulf Stream, the Bay of Biscay currents, and the prevailing winds. Its forward progress averaged an amazing 3.3 miles per day. Since that early time, the whole Odyssey has given me a unique lifetime story.

INVITATION TO HELL
Leo Hyatt

FOUR SLEDGEHAMMER BLOWS SHOOK the plane and caused a near-total blackout for me, as I felt a spinning sensation and tried to correct it. I do not remember my ejection from the bird. My left forearm was broken by the seat, and since there was command ejection when the front seat fired, my Radar Navigator, Wayne, left the plane just before I did. When I came to, I noticed it was very quiet. I looked up, saw the chute, and the ground was coming up fast. I landed in a bunch of scrub brush.

It was a struggle getting out of the chute harness, helmet, and oxygen mask as my left arm had been shattered at the shoulder socket. I got up and started running, leaving my gear behind—didn't know where I was going, as we were over 110 miles from the coast and chances of pickup were remote. It looked like a thousand people were after me.

They were shooting, and I noticed the bullets went *"snap-pop"* as they went by. One hit me in the right arm and down I went into a clump of bushes. I took out my .38 service revolver and threw it away, along with the two radios, after trying to make a call that we were on the ground. I couldn't fight an army with six tracer rounds.

We (Wayne and I) were captured by a group of soldiers who took us to a nearby village and held us in a cave overnight. A medic came and dug the bullet out of my arm as well as a nail from my boot from my right heel. He put my left arm in a sling and left.

The next day, our captors moved us by truck and Russian biplane to our new home, the Hao Lo prison, better known as the Hanoi Hilton.

I was put in a room with a table, chair, stool, slop bucket, teapot, and cup, stripped to my shorts, and given a pair of red and gray striped pajamas, black cotton T-shirt, and shorts, and a pair of flip-flop sandals made by cutting the sole shape out of a tire with pieces of an inner tube across the instep to hold them on (we called them zaps).

I had time to take stock of myself before the inevitable happened. My left shoulder looked to be about four to five inches out of the socket. The way it hurt, it had to be broken in many places, and the upper arm was broken about two inches below the ball. Everything else was minor in comparison to that.

The door opened and in walked the devil himself in the shape of a small Vietnamese man in uniform who was about to give me my introduction to Hell. This English-speaking (but not very well) officer sat behind the desk, motioning me to sit on the stool while an enlisted man stood behind me. The interrogator asked my name which I had to spell, and he said, "You Vietnamese name Hy." Then came rank and serial number, and I was thinking this was just as it was supposed to be, according to survival school training.

Then came a series of questions like what squadron, ship, type of aircraft, etc. I was from, to which I replied I couldn't and didn't have to answer those types of questions according to the Geneva Convention. This really pissed him off. He started yelling at me that I was a Black criminal, an air pirate, there was no Geneva Convention, there was no declared war, it was an unjust war, and I had a very bad attitude for which, "You must be punish!" This was the first of countless times I was to hear this. I was grabbed from behind and then it began.

The wrists were bound together behind my back by rope and the ankles were secured spread eagle by shackles to a rusty metal bar across the top of the feet. The rope was looped around the tips of the elbows which were pulled together and then the rope passed over a shoulder to the bar. The whole thing was cinched down to fold me up into a suitcase position—arms extended to dislocate the shoulders, (but my left one was already detached!) and the legs pulled up to form an unbearably painful position wherein it was nearly impossible to breathe. Screaming got you a filthy rag in the mouth, held in place by a piece of rusty pipe.

There can be no way to describe the pain or the position you were in. The enlisted guy doing the work was enjoying it, and it must be said they were masters at inflicting massive pain for prolonged periods of time. They would leave for a smoke break, while parts of you went numb and you gasped for air and wondered how in God's name you could survive this. I remember asking God to give me the strength to make it for a minute, count it out, and then ask for one more. This ritual went on and on, during the next two to three weeks.

They would come back in, release the rope, and the blood rushing back into the extremities and the nerves coming back online was as painful as the tie-up.

"You answer the question now."

"No."

Back in the position. I honestly do not know the number of times this went on over many days and nights. It was terribly hot, and I could not get enough liquid in my body with two liters of boiled water flavored with a bit of tea and two bowls of cabbage soup each day.

I couldn't eat much as I was nauseated. He was getting more

upset with me and yelled at the goon doing the damage one day. He cinched down extra hard as he kicked me in the right rib cage. I felt the ribs crack and my neck broke around the fifth vertebrae (confirmed by X-ray after I got home). I started to vomit and had to blow it out my nose as I had a rag in my mouth. The English speaker was down on the floor next to my face and kept yelling, "You talk now?" I figured this was a pretty stupid way to die—drowning in your own vomit—so somehow, I indicated I had had enough.

I was turned loose and refused to answer anything until I had crawled back up on the stool. The main purpose of the questions now was: what would future targets be?

Of course, I did not have a clue, but figured I had to give him something. I said I needed some maps, and he ran out to get some. I then proceeded to point out various areas all over North Vietnam, saying that there was a truck park, rail siding, someplace along a road, etc. This went on for several days and each time I would give him something, he'd run out to another room, and I could hear him yelling as if on a very poor circuit phone connection. If they sent flak sites, SAMs, or whatever to the places I told them, they burned up one helluva lot of fuel.

The next step was to write about myself in a sort of biography. I was put into another room and told to write. I would jot down a few sentences about nonsense and then write that I had much pain and had to rest. I did a lot of crying, feeling sorry for myself, along with a bunch of "Why me, God?" I had broken faith with my fellow prisoners by giving more than name, rank, serial number, and date of birth in accordance with the Code of Conduct.

I was a total and disgusting excuse for an officer in the Navy. What would the rest of the POWs think? I found out sometime later, after communication links had been established with the senior officers in the camp, that the rules were to not endure torture to a point of no return but to give something and live to come back and fight another day.

We were most fortunate to have a superior group of seniors

who had some great advice, made tremendous policy, and kept us together in a unified front. There were a few POWs who strayed, doing anything to save their own hides, and eight were given early release because of propaganda they made for the V.

After some time, punctuated by intermittent interrogations and torture sessions, my injuries were becoming grave. Finally, unable to leave my bed unassisted, I began awaiting death. Then help arrived in the form of an Air Force pilot, also injured, but still able to help me. His nursing care essentially saved my life.

Time slowly passed until about October 1967, when Ed and I were split up (he died later, still in captivity). I went to the Zoo Annex to a room with five other guys. One had been there for enough time to learn the ropes of how to communicate cell to cell through the walls, by using a tap code or flashing pot lids between buildings, or by setting up drops in the bath areas. What a blessing that was as we were now in the comm link and could find out what was going on.

The main technique for communicating was the five-by-five system so we could tap code through the walls. It was based on an old prisoner code and made up as follows:

$$\begin{array}{ccccc} A & B & C & D & E \\ F & G & H & I & J \\ L & M & N & O & P \\ Q & R & S & T & U \\ V & W & X & Y & Z \end{array}$$

The letter K was eliminated, and C was used for it. The rows were numbered 1 through 5 as were the columns; hence, "Any news" was tapped out with 1 tap, 1 tap, pause, 3 taps, 3 taps, pause, 5 taps, 4 taps, pause, 3, 3, pause, 1, 5, pause, 5, 2, pause, 4, 3. Many hours were spent tapping on the wall to the guys next door and the calluses on the knuckles of the index and middle fingers grew—they are still there to this day!

Christmases came and went, surely the saddest time of year for us. Our captors used Christmas as an opportunity to generate propaganda.

The third Christmas, I was told to suit up for a trip to the "Head Shed." Word came through the walls that guys being called to the interrogation rooms were being given packages sent by our families. So, the word was, go get the goodies and do your best to not make propaganda for the V. In other words, don't smile for the cameras! Off I went, spruced up in a set of striped pj's, having just been made to shave, and gotten a haircut from the guard.

Sure enough, lots of photographers were present with lots of lights. I sat on the customary stool and a package was placed in front of me. "We show you humane and lenient treatment to have a Christmas package from home" I had been sitting with my arms folded, having slowly and with a show of pain lifted my left arm into place, all the while extending the middle finger of each hand in the classic "bird" position.

I was told to look happy and smile—which of course I wasn't—with the cameras flashing the whole time. They didn't like what was going on, so I was told to get my stuff and go to my room. The photo in *Life* magazine a short time later was the first that my family knew I was a POW and removed my name from the MIA list. Intelligence people called my wife to ask if she had seen the photo and if she had seen anything. She said it was obvious I was totally pissed off, and I was shooting the bird at the whole world. The intelligence folks had not noticed!

Possibly the biggest blow to morale was the cessation of the bombing. The prisoners felt abandoned. This was exacerbated by the Hanoi visits of American anti-war activists, the most prominent being Jane Fonda.

We lived in constant fear of being pressured or tortured to go meet one of these delegations or to write letters to various Congressmen supporting their views on the immorality of the war and how we agreed with them. Quite a few of the POWs had to endure severe torture until they agreed to meet with these Americans we all felt were traitors. Several of the men tried to slip small pieces of paper with their names on them to these people. In at least one case, that individual promptly handed over the papers to the English speaker with the delegation, and the guys were, of course, punished when they returned to their cells.

I believe we all agreed that people had the right to speak their minds about the war, but—and it is a mighty big BUT—when it is done in a manner that (a) aids and abets the enemy, (b) prolongs the war by giving him hope they will soon win because the USA will be forced by the movement to quit the scene, and (c) puts American fighting men who are held captive in grievous harm's way, IT IS ABSOLUTELY wrong and traitorous. Most of us would agree it prolonged the war. I have the scars to harbor my hatred for these people, and I suppose I will take it to my grave. I am proud to have been an Air Pirate and the Blackest of criminals because I am honored to have lived so long with some of the finest, bravest men this country has ever produced.

If we had turned our backs on the Vietnamese, we would have been the biggest hypocrites on earth. It was simply a matter of helping them to become freer and more democratic, and the first order of business was to keep them from falling under the Communist umbrella.

We lived in limbo until December 18, 1972, when we noticed something out of the ordinary just before bedtime. A noise such as we'd never heard before. It was as if lots of jet aircraft were a long way off. Air raid sirens went off and the lights went out and someone yelled, "Holy s--t! That's high-flying B-52s!" Someone else said, "We're getting out of this f-----g place now!" And then it started.

The thump and bang of anti-aircraft guns joined in with the launching of surface-to-air (SAM) missiles. The drone of the B-52s became louder and then the bombs began to hit. They hit ammunition dumps, POL (petroleum, oil, lubrication) storage areas, warehouses, power plants, rail yards, etc., and the whole world seemed on fire. The night became day. The ground shook, and it was an awesome sight and sound display. It was the most wonderful, exhilarating thing any of us had ever seen! WE WERE GOING TO GO HOME! The good ole USA had come to get us out.

No city or nation could withstand the onslaught of hundreds of B-52 aircraft in an unending stream for long. And, as we learned from the newly shot-down airmen, this was occurring in the port city of Haiphong and other major areas all over North Vietnam. This continued until just before Christmas and then stopped.

After a few days, a major move was underway again. Our most senior officers were called to the headquarters building. They were told to control their POW personnel, make everyone behave, and the treatment would be relaxed. Word came out that there were talks going on in Paris regarding what was to happen, and that we would be going home soon. Our SRO (Senior Ranking Officer) asked that the sick and wounded go home first, to be followed in order of shot down, longest tenure first, or none of us would agree to go.

In a few days, we were all called together in the main courtyard. There were a lot of cameras and a lot of brass in attendance.

An English speaker read about the Paris Peace Accords that had just been signed and said, "The humane and lenient people of the Democratic Republic of Vietnam now allow you to go home to your families," or words to that effect. All of us stood there stone-faced, showing no emotion. Silence engulfed the courtyard for several minutes. The English speaker waved his arms up and down, and said over and over, "It okay, be happy, you go home." We gave him nothing and the cameras got nothing. After several minutes we were told to return to our room. Then, I can tell you, all hell broke loose!

The day finally came when we were issued a small overnight bag, a pair of pants, a shirt, a windbreaker, socks, and black shoes. We were told to get dressed in our new duds and load some buses pulled into the courtyard.

We moved slowly through the city, and it was a sight to behold. It was devastated—piles of rubble everywhere. We arrived at the airport ramp area. A long table had been set up, behind which sat some of the English speakers who previously had conducted the interrogations and torture sessions.

But the most beautiful sight in all the world was parked on the apron area with its back ramp lowered and a large American flag painted on its tail. There never has been nor will there ever be a more wonderful, beautiful sight than that C-141 airplane and the American Flag on its tail—enough to make grown men cry from sheer joy and most of us had tears rolling down our faces.

We were met on the ramp by four of the most beautiful, best-smelling, US Air Force uniformed women God had ever created, who took us to a seat and gave us fruit juice. We felt we had died and gone to Heaven!

As we took off, the cabin was strangely quiet. I guess we suspected it was all a joke, we would have to go back and land, or they would shoot us out of the sky as we climbed out toward the Gulf of Tonkin. Then the pilot announced we were "feet wet" (over the ocean)! Now the cabin erupted with all sorts of yells, hoots, hollers, hugging, etc. We were finally out of Hell and on our way home!

I have been asked many times what made it possible for me to survive. There are several reasons. First, I had a tremendous hatred for my captors, their inhumanity, brutality, the whole Communistic crock of lies, etc.

Secondly, although I felt that my God had forsaken me (part of my feeling sorry for myself and my predicament), I never knew there was a large group of people praying for me every day. At the church where I had grown up, a tree had been planted on my behalf to remind parishioners to keep it up. It must have done a lot of good because I was given the strength to survive—I didn't do it all by myself!

Thirdly, I must go back to my high school football and basketball coach who had instilled in me an unshakeable ethic of "get up when you get knocked down and have the guts to face the bigger, faster, stronger guy across from you."

Last, but not least were the important principles of leadership, love of country, honor, and duty that had been drilled into me while I was at the US Naval Academy. You simply didn't quit, and it was imperative that one kept faith with your fellow officers and the men with whom you served.

We each went through a press conference and an awards ceremony over the next several weeks, at which I was awarded of many decorations/medals. The two of which I am most proud are the Silver Star and the Bronze Star with V for valor.

A NAVY JUNIOR'S FAR-EAST ODYSSEY, 1938–1940
Bob Strange

IN 1937, MY DAD (R. O. Strange, USNA Class of 1928) was ordered to USS *Marblehead*, a light-cruiser, home-ported in San Diego. Soon after reporting aboard with the increased threat to peace in the Far East, the decision was made to relocate the ship to the Asiatic Fleet with Shanghai, China, as her homeport. Initially, dependents were not permitted in the area, but this restriction soon was relaxed providing that suitable housing could be found. Japan, by this time, had invaded parts of China and was looking for any excuse for making further inroads into the country.

My mother and I traveled to the Far East on a commercial ship in 1938 and took up residence along with other Navy families in the Cathay Mansions within the International Settlement of Shanghai. Meanwhile, the ships of the Asiatic Fleet would conduct maneuvers throughout the region and visit ports of call in "Show the Flag" missions. There had been some incidents with the Japanese, the uppermost one being the sinking of the Panay in

1937, causing an increase in readiness by US Forces to be implemented in the area.

During the summer months, the fleet would relocate to areas with milder weather conditions and would proceed to Tsingtao in northern China on the Shantung Peninsula. Dependents would follow the ships, and we took a coastal steamer to Tsingtao where we found a suitable place to live in a German boarding house. It was located within a compound surrounded by high walls and in the center of the compound, we found the German Swastika flag flying.

I became friends with a German boy who lived in the compound, and once each week, he would show up to play dressed in his Hitler Youth uniform. The Japanese maintained forces on the Shantung Peninsula, but there had not been any recent incidents. One day, while I was playing in the compound, I found someone had left the gate to the outside ajar. Being an adventurous type, I ventured out only to discover after a while, the gate attendant had locked the gate. At about this time, I became aware of a contingent of Japanese soldiers in full battle gear performing skirmishes in the street adjoining the compound. As a six-year-old, I was petrified and soon got the attention of the gate attendant, who let me back into the compound.

In the winter months, the ships relocated to warmer waters, and we found an appropriate apartment in Manila, Philippines, near the Fleet Landing. While living in Manila, I can recall going on family excursions to Baguio in the mountainous region of Luzon, where the US military-maintained cottages for rest and relaxation. I had been told that indigenous natives who lived in the region were headhunters, and during our stay, we visited a nearby native village and observed shrunken heads on display about the size of an orange. These certainly made an impression on a seven-year-old.

Returning to Shanghai, we again took up residence in our apartment in the Cathay Mansions. By this time, the Japanese were bombing the city on a regular basis and the decision was made that dependents should make plans to return to the United States.

We departed Shanghai in March 1940 aboard a commercial ship with many other dependents to travel back to the United States and home. It was time to get the heck out of there.

LONG CARRIER DEPLOYMENT
Frank Alvarez

I FIRST THOUGHT TO WRITE about my second Western Pacific Cruise as the Chief Engineer of the USS *Midway* (CV-41) after observing the attention the *Lincoln* (CVN-72) received upon her return after a nine-month deployment. She had just set the record for the longest carrier deployment since the Vietnam War, and the longest deployment ever for a nuclear-powered carrier. I am sure the ship's company and air wing deserve all the kudos they received.

However, I don't think anyone else except our immediate families was aware of our return as we slid into our berth at Alameda Naval Air Station on March 3, 1973, after setting the record of eleven months for the longest deployment for a carrier. The public reaction to the military in those days was grim and thinking about it made me a little sad. Particularly, since we had a very arduous deployment, with eleven brave aviation personnel dead or lost at sea, seven aviators missing in action, and six more prisoners of war at that time. *Lincoln* lost none, and I don't think it was just because of their excellent training as implied by the press.

This history story starts on April 5, 1972, as the attack carrier USS *Midway* (CVA-41) was sailing off the coast of California into the first several days of operational training which were supposed to last a couple of weeks. I was the Chief Engineer of this conventionally powered ship. Unexpectedly, we received orders to return immediately to Alameda, our homeport. Immediately, rumors started flying, the most popular being that we would deploy early to Vietnam waters. Our current date for our next deployment was around June 1, and we needed those seven weeks to complete all our training. That rumor proved to be correct however when on April 7, the Chief of Naval Operations, Admiral Zumwalt, flew out to the Alameda Naval Air Station and gave us the news we were needed now to help respond to a North Vietnamese massive offensive across the Cambodian border. We were to sail in just three days.

Right on schedule, we sailed under the Golden Gate Bridge on the tenth of April with no idea when we would return. Soon thereafter, Carrier Air Wing Five landed on board. Just nineteen days later, on the thirtieth of April, since we were crossing meridians to the west, the air wing made its first strikes over Vietnam. Ships have a life of their own, and *Midway* always had an outstanding reputation since her commissioning in 1945. We all knew this would be a very successful deployment even though we did not complete our training and had only three days to load out. Many in the crew and air wing did not have time to square away all their personal business before departure.

We soon settled into a routine of long periods on Yankee Station with short visits to Subic Bay in the Philippines for some needed material upkeep. While on station, we would launch aircraft every day, and replenish for about five hours every other night, refueling, taking on aviation gas, replenishing bombs, and loading new stores. Since *Midway* was multi-compartmented due to the WWII experience with the loss of the USS *Franklin* (CV-13), mine was the largest department on board with approximately seven hundred enlisted and eighteen officers. This number was necessary to man all the compartments under engineering responsibility, stand the required watches, and perform the necessary maintenance and repairs throughout the ship.

While the aviators were putting their lives on the line every day, the best thing we could do in engineering was to maintain the ship in tip-top material condition. This made it possible to always make enough speed to provide sufficient wind over the deck for safe flight operations, provide enough reliable steam to the catapults for hot shots every time, keep the arresting gear machinery in excellent working order, ensure good working aircraft and weapons elevators, maintain well air conditioning spaces, keep the gallery equipment in proper working order so excellent meals could be prepared and provide safety in the way of firefighting and damage control. This we did outstandingly. The only times we had any discussions with the air wing was when we insisted the ready rooms be furnished and enhanced using only fire-retardant materials, water be conserved as much as possible as the air wing washed down their planes each day, and some of the air wing and ship's company personnel took non-Navy showers. Incidentally, classmate John Disher was the Executive Officer of VF 151 of the attached air wing.

An unexpected kudo for my department arrived by way of a letter to me from the Assistant Ship Material Officer on the COMNAVAIRPAC staff. It appeared *Midway* would be making port in Singapore soon, and he wanted me to investigate any possible capability for carrier repair work there. He ended the letter with the following: "In the past month, I briefed VADM Holloway (the new COMSEVENTHFLT) and RADM Flanagan (the new COMCARDIVONE whose flag was on *Midway*) as they toured through the staff prior to assuming their present jobs. I advised both that, in terms of material condition, I felt *Midway* was our best carrier. "Keep up the good work!"

It was necessary we kept *Midway* in the very best material condition because, during the entire deployment, we were always kept in the hottest spot. The above-referenced letter mentioned an impending visit to Singapore was dated July 18. We finally got there for Christmas. This was only possible after the unrestricted bombing of North Vietnam, which was allowed by the President earlier in December, finally brought the enemy to the bargaining table. During the deployment, we eventually stood by at one time or another for every carrier out there due to material problems except for *Hancock (CV-19)*. She had a wooden flight deck and could not take on the assignments of the steel-decked carriers. *Kitty Hawk (CV-63)* didn't have a material problem, but she eventually had to stand down due to personnel problems.

The diversion of a port call in Singapore was greatly appreciated by the whole crew. Not only did it take place over Christmas, but also it turned out to be memorable because Bob Hope and company came on board and put on a wonderful show. I personally felt fortunate, since that was the second time a Bob Hope show was performed on my ship during seven WESPAC deployments. He had also come aboard the cruiser *Los Angeles* (CA-135) in Buckner Bay, Okinawa, just before Christmas 1958. On that occasion, young officers had the opportunity to meet and talk to Jayne Mansfield in the wardroom. That was special!

All our successes during this deployment came with a high price for the air wing. Eleven members died or were lost at sea. Seven aviators were missing in action, and six more were prisoners of war. I do not know the final disposition of the last two groups since the air wing left the ship upon our return.

USS *Midway* (CVA-41) and Attack Carrier Air Wing Five did receive outstanding recognition for all our accomplishments in that we were jointly awarded the Presidential Unit Citation for the period April 30, 1972, to February 9, 1973.

Soon thereafter, we were finally released, and we returned to Alameda on March 3, 1973. That was just one week short of eleven months of deployment. At that time, that was the longest deployment for a carrier. I am not sure, but I think it still holds true as the war started to wind down after that until the end of 1975. Incidentally, my Meritorious Service Medal Citation signed by

COMSEVENTHFLT states in part, "Met all operational commitments *during a record-setting number of combat days* on an extremely arduous deployment." That is an even better record to have than the one above. Likewise, I don't know if that record still stands.

As I mentioned earlier, ships seem to have a life of their own. *Midway* always had an outstanding reputation. She was the best ship in which I ever served!

MY CAREER WAS OVER BEFORE IT STARTED
Sam Underhill

I WAS ASSIGNED to the USS *ROWE* (DD-594) for my senior-year cruise with the great liberty (free time ashore) ports of London and Copenhagen on the schedule. I was surprised to be selected as the Midshipman Executive Officer, especially since I was a few hours late returning from London liberty ashore. (Missed the last train but hitchhiked back in a delightful truck carrying live animals.) This assignment gave me more exposure to the Ship Officer's wardroom than I wanted, in trouble.

The Commanding Officer, whose name I have conveniently forgotten, was unequivocally the wildest I've ever seen. You could rarely anticipate what he would say or how he would maneuver the ship. His favorite pastime was kicking the Squadron Commander (his superior), who had just been selected for Admiral, off the bridge when there was a disagreement, which happened rather frequently!

But on liberty, he was a real wild man. While I was not on liberty with him in London or Copenhagen, his reputation was on full display ashore in Cuba. As the officer's liberty was one hour later than ours, we typically searched out the wardroom officers to seek permission from the Midshipman Officer Supervisor, LCDR Traynor from Annapolis, for an extension to midnight. He always accommodated.

My first ride back to the ship on the last bus with the officers was a real eye-opener. The Commanding Officer ordered all of us to line up outside the bus door as side boys and to catch him when he jumped out. We all closed in tight but damn if he didn't jump beyond us. Dirty and bruised, he got up swearing like the sailor he was, threatening us with a loss of liberty if we didn't catch him the second time. Every night thereafter, we never missed.

As midnight approached the last night, we must have been discussing what sports we did at Annapolis, because someone (I'd love to know who) mentioned that I wrestled. That got the CO's immediate attention. Upon confirming I was on the wrestling team, he shouted, "Let's wrestle." I almost panicked as he was the CO, twenty-five to thirty pounds heavier, the ground was nothing but hard top, and I didn't know if he was serious. I quickly learned though when he grabbed me in a headlock and started throwing me around. The officers unsuccessfully tried to stop him. I looked forlornly at Mr. Traynor, but he couldn't help either.

At this point, I only had two options: let him throw me around and possibly get seriously hurt or protect myself. So, I went under him, took him down, put him on his back, threw on a Half Nelson, and held on as he squirmed mightily to get out. After three to four minutes, he was exhausted, and the officers convinced him I had won fairly.

One of my arms was scraped and bleeding and my trousers were torn, but other than being extremely concerned I was fine. There were no side boys that night, but the officers did have to put the CO in his sea cabin. When we got underway the next morning, however, the CO did not appear on the bridge or leave his cabin for several *days* because his back was so torn up! Mercifully, no one in the wardroom ever mentioned what had happened for the rest of the cruise. Nonetheless, I was petrified my career was over before that cruise was even completed despite assurances from Mr. Traynor. I was convinced I had wrestled away my career in one of my best matches ever! Fortunately, it was not so.

[*Compiler:* Way to go, Sam!]

PEARL HARBOR, THE BEGINNING OF AN ODYSSEY
Tom Marnane

Waiting for the Bus

I WAS A YOUNG BOY seven years and four months old, standing in front of my house at Wheeler Army Airfield, Oahu, Hawaii, on December 7, 1941, at 8:00 a.m. I was waiting for the bus to Sunday school, and I was wearing my little boy's Army uniform like my dad's.

Wheeler Field was busy on the weekends because the Army reserve officer aviators came out to get their flight time. I didn't know anything unique was going on until airplanes with Rising Sun insignia started flying close overhead down my street (four houses from the airfield). I knew something was happening when I heard airplane machine gun fire and spent shell casings falling on the street. I immediately started filling my pockets with the casings and ran up the street to see what all the smoke was about at the airfield (we kids frequently went to the field and talked to

Aftermath of the Japanese attack on Pearl Harbor on December 7, 1941.

pilots and got to sit in cockpits and get fed whatever was available in the ready room). It was about this time that my dad, West Point, Class of 1931, came running out, grabbed me, and took me back to the house where we stayed (mother, brother, and myself) for several hours, while my dad went to report in. When he returned, he was wearing a gun and in combat fatigues (no one knew if there were to be troop landings or other attacks).

The next day, my mother, brother, and I were evacuated to a school on a pineapple plantation where we stayed for four days. Large gymnasium, cots to sleep on, lots of kids, hamburgers, ice cream, pancakes, hot dogs, and movies—it was great fun, although I doubt my mother thought so!

We returned to a set of quarters (not our own) where we observed blackout rules and huddled under furniture for a few days. Then we returned to our house, which we found was riddled with bullet holes, and had a gun emplacement and soldiers in our yard. We set about getting ready to evacuate. My dad drove us around the island, and we saw the damaged ships, a captured two-man submarine that had gone aground, the remains of a Japanese airplane that had been shot down, and the still-smoldering mess of airplanes and equipment at Wheeler Field (think: *Tora! Tora! Tora!* movie). On Christmas Day, December 25, 1941, we were evacuated to the US on the SS *Matsonia*, a Matson Lines passenger ship. I remember it well because Christmas dinner was peanut butter and jelly sandwiches and warm soft drinks while we waited to board. We sailed that day under convoy, arrived at San Francisco eight days later, and proceeded to Tulsa, Oklahoma, to stay with my grandparents for the next four years.

After the War in the Pacific, the Army of Occupation in Japan

My dad returned to us on December 7, 1945. He had gone from Captain, USA to Colonel, USA during the war, was highly decorated (Twenty-Fifth Infantry Division), had had "every kind of jungle rot known to mankind," and couldn't get enough blueberries and strawberries. Within three months, he was ordered to Germany to work on the final cleanup of Nazi Germany.

My family returned to Tulsa. My dad returned a year later, we had a couple of short duty assignments in Atlanta and Chicago, and my dad was ordered in 1949 to Japan with the Eighth Army in the Army of Occupation. We accompanied him. At the age of fifteen, I had a lot of different emotions going to the "house of the enemy," the hated "Japs" (my grandfather called for "silence on deck" as we listened to the news as a family every night during the war and read aloud the almost daily censored letter from my dad).

We lived (with nine servants—jobs created by the US to provide work for destitute Japanese) in a Shell Oil Company executive's house (on a hill overlooking miles of slums) in Yokohama. I spent my junior year at Yokohama American High School, played football, worked on the newspaper, and generally had a good year but never got completely comfortable with the Japanese. I had little interface with the Japanese except for the servants.

General MacArthur was a guest for cocktails in our home, and I was my dad's bartender. We rode Army Special Services' private train cars twice a week each time sightseeing in Japan. Among other places we visited were the Japanese Naval Academy, Hiroshima and Nagasaki (still no reconstruction—unbelievable sites and horror), and Tokyo's Ginza, now a major shopping street, with hundreds of shanties selling goods. On June 25, 1950, North Korea—with Chinese support—attacked South Korea. Two days later, my dad was ordered, with the Eighth Army, to prepare for the landings at Inchon. My family returned to Tulsa on USNS *Patch* in 1950, and I delivered my first talk on the attack of Pearl Harbor to my Tulsa High School Class of a thousand students.

Pearl Harbor and Myself in Active Navy Years and After

As Puget Sound Naval Shipyard Production Officer, the USS *Missouri* (BB-63) was mothballed in my front yard. We conducted promotions, retirements, memorials, and other ceremonies on her quarterdeck. She went to sea (under tug power) every two years to turn her around so one side didn't get all of the sun—I went along so I could say I had been to sea on all four *Iowa* Class Battleships (three on Midshipman cruises). She is now at Pearl Harbor.

In 1980, I was ordered to command the Pearl Harbor Naval Shipyard—untouched on December 7 and key to repairs during the war years. In front of the shipyard office building is a miniature Japanese submarine from December 7th. The office building is a National Historic Site. Under my office in the building were several basement rooms where the Japanese code was broken in 1942 by the tedious efforts of out-of-work bandsmen from the damaged ships. I presided over the dedication of those rooms as a Historic Location. I had many opportunities to conduct ceremonies at the USS *Arizona* Memorial (what a pleasure and honor) and was responsible for the maintenance and upkeep of the tourist motorboats to and from the Arizona Memorial (heard about it quickly if any were not functioning or broken down), and I talked to groups of tourists on occasion about the attack on Pearl Harbor.

USS *Coral Sea* (CV-43) had a collision at sea—busted up the bow and it had to be replaced. The shipyard turned to (went to work), still working with experienced WWII workers, built a new bow, and had her out of there fourteen days after she came in for repairs. When Japanese Navy ships came in for repairs, different emotions passed through my mind hearing the Japanese National Anthem played in the mornings. We entertained and were entertained by Japanese officers and Midshipmen. Charlotte, my wife, and I made two trips to Japan while there—wonderful experiences.

After Retirement from the Navy

I got a job with Bechtel Corporation in San Francisco after the Navy as a Naval Architect for Bechtel, Civil, and Minerals. My first assignment was to design a floating runway sufficient in size to handle aircraft training flights in and out of Yokosuka to cut down on sound complaints in neighboring communities. It never got built, but young Japanese engineers from Kuma Gai Gumi Corporation worked with us—the hardest workers I have ever been with. I had to turn out lights to make them stop at night.

I changed jobs in 1987 and became VP of Logistics for Matson Navigation Company. I handled everything except ships, although I did due diligence inspections before leasing them. I visited my terminals up and down the West Coast and in Hawaii once a month. Terminals in Hawaii were at Pearl Harbor, Maui, the Big Island, and Kauai. SS *Matsonia,* a WWII passenger ship, is now a container ship. I took visiting friends on Pearl Harbor guided tours with WWII emphasis.

Now, as a retiree, I address schools and various groups on request about my experiences as a seven-year-old survivor of Pearl Harbor.

Bottom Line

I have never completely gotten my Pearl Harbor and Pacific

experiences out of my mind—not traumatized, just mindful. I still have trouble reconciling how I feel about Japan, Germany, North Korea, and China for separating me from my dad for six of my formative years—and for no aggression on our part. I have worked many times with Japanese, German, Korean, and Chinese people designing and purchasing equipment at Bechtel and Matson and visiting each country—and they have all been good experiences. I played the role of North Korea during the War Games at the Naval War College.

My dad died on December 7, 1990, and was buried at Arlington National Cemetery with full honors (even the riderless horse with boots turned backward—he had been a cavalry officer with his own horse and "Batman" in the 1930s). He never wanted to return to Hawaii or anywhere in Asia.

POSITIVE LEADERSHIP
Jim Beatty

ONE DAY IN NOVEMBER, during my First Class (senior) year, and a week or so before the Army/Navy football game, I returned to my room very unhappy that I had blown a quiz. I angrily slammed our door open and, shortly thereafter, heard shattering glass. What to do? I didn't want a report chit (demerits) that might keep me from going to Philadelphia and the big game. Well, because it was close to the big game and the Plebes had started to decorate, I decided to make the best of a bad situation. Somewhere in Bancroft Hall, I found some cardboard, some thumbtacks, and tape, and proceeded to build an Army Mule Meat Market. I put an awning above the now-open-door window and a counter at the base. I filled some khaki socks with paper, wrapped them in "Beat Army" stickers, and strung them up like sausages. I moved the "in charge" of the room tag to my name (so that my roommates would not be held responsible for any trouble) and hoped for the best.

It wasn't long before our Company Officer, Captain Patton

(Yep, the son of the famous WWII General), visited our room as part of his normal rounds. The conversation went something like:

"Who built this?"

"I did, sir."

"Very funny, Beatty."

I waited for the other shoe to drop.

Saturday morning arrived and I boarded the bus for the big game adventure. Maybe he would hold the report chit until next week, and I could work it off before Christmas leave!

The game was a tie, the parties were fun, the weekend was great, but reality set in, and we were back in Annapolis Sunday evening. I left the Army Mule Meat Market in place, figuring I could wait until Monday for the inevitable.

Upon my return from class Monday morning, I was greeted by a Bancroft Hall maintenance man who informed me that if I took down my display, he would install a new pane of glass. As you can imagine, the "Market" disappeared in a hurry and suddenly our room was back to normal.

To this day, I am convinced Captain Patton had the maintenance man camped outside our door with replacement glass in hand, waiting for my return, letting me off the hook, but once again exercising very positive leadership.

REPORTING ABOARD
Urb Lamay

IN EARLY AUGUST OF 1957, I ARRIVED in Coronado, California, with my bride of several weeks, and upward of twenty pounds, having eaten very well since graduation from Annapolis in June and especially during the immediately preceding eight-day cross-country trip. Former company mates Cleve Loman and Dick Enkeboll and their wives helped us find suitable housing and a few days later, the newlyweds were at Lindbergh Field. Ensign Lamay (me) was ready to embark on the first leg of what was to be about a three-week comedy of getting to my ship, USS *Braine* (DD-630), which was then in the Western Pacific.

But first, a picture of this young ensign in his khakis—a quick check to make sure the cap was not too rakish, a smile (hardly genuine since I was about to leave a teary bride), then brace up for this picture. The latter action was a bit more than the buttons on the now-somewhat-undersized khaki blouse could withstand.

So *pop, pop, pop* they went.

The tears on my bride's face were instantly transformed into a burst of almost uncontrollable laughter—not exactly a bad thing at that time. (A damn good thing I didn't bend over too quickly to retrieve those buttons and test the strength of my trouser seat seam.)

A few days in San Francisco followed, getting necessary shots and just waiting for the call to board a bus to Travis AFB for my flight across the big pond to find my ship. Of course, I used this time wisely; I spent nearly all of it learning to sew buttons on my blouse and purchased a nonregulation brown leather belt—a little more accommodating to my present circumstances than my June and earlier-sized khaki web belt straining around my middle. I was seated on sagging canvas seats mounted so that one (me) faced inward. Our "head" (toilet) on those long flights was a noisy funnel device. After we finally got off the ground on whatever it was, I flew on.

The trip was uneventful, uninteresting, and seemingly unending. Nine hours to Pearl Harbor, nine hours on hold there, then another nine hours to some mid-Pacific Island for refueling, and finally nine hours to Tokyo where I was stashed in barracks for several days. Checking in several times a day with the Navy Fleet Operations Center at Yokosuka, Japan, became the routine as that office first tried to find my ship, *the Braine*, and then arrange for transportation thereto.

At last, the day and hour arrived, and I found myself on the way to Taipei, Taiwan, and thence to Kaohsiung to meet the ship. But first, a refueling stop at Okinawa. Upon arrival there, the passengers were informed they would have an hour to get lunch if they wanted to, either at an enlisted men's mess hall or at the Officer's Club.

Leaving everything on the plane, except what I was wearing, which included my new leather-belted khakis, I was one of the several transients chosen for the Officer's Club. As I recall, I had a decent meal there while periodically checking my watch to make sure I returned to the plane on time. But when one such check divulged the exact same time as the previous check, I bolted from the table, out the door of the Officer's Club, and before getting very far was treated to the sight of my plane picking up speed on the runway and lifting into the air.

I didn't remember having been trained at USNA for such a contingency and so had to draw on other resources—which at about that time, seemed very lean indeed. Finding my way to Flight Operations, I told personnel of my sorry plight and pondered anxiously as to what kind of a holding cell they would put me in. I was told to check in at the Bachelor Officer's Quarters and to check back with them twice a day until they had arranged further passage. With fast-depleting cash, no checkbook nor credit cards in those days, I purchased a few necessary toiletries and began the seemingly interminable wait for "further passage."

Two days later, wearing the same, and now none-too-presentable clothes as when I departed, I was once again on the way to Taipei. Arriving in the dark, I was directed to a ground transportation office—if one chose to call it that. It was a shacky-looking screened-in hut whose two or three bare unshaded light bulbs attracted every sort of insect imaginable all to the delight of the hundreds of chameleon lizards. They were in addition to the two Taiwanese men manning this seemingly "jungle outpost."

How was I to be reunited with my luggage and khaki blouse which I had left on the plane in Okinawa? That's later. First, I had to report aboard the ship I had been chasing all over the Pacific. I, without my blouse on to mask my nonregulation leather belted trousers, sheepishly approached the deck of the *Braine* with seemingly a thousand faces of officers and crew watching every step this new officer made as he reported aboard. With as firm a "Permission to come aboard, sir?" as I had ever given, I boarded my ship and ended the saga of the first several weeks of my naval career. What a way to start!

PEARL HARBOR ATTACK AND RESCUE AT SEA
Bob Strange

FOLLOWING USNA graduation, I reported to my first ship, a Fletcher Class Destroyer, USS *Jenkins* (DDE-447) out of Pearl Harbor and was duly assigned as the prospective Anti-Submarine Warfare Officer along with a host of other junior officer collateral duties. Deploying to the Western Pacific soon after my arrival, we spent six months operating with the Seventh Fleet as a carrier escort ship and conducted numerous anti-submarine warfare exercises.

Returning to Pearl Harbor after that deployment, I had nearly achieved my Officer of the Deck (OOD) underway qualifications and soon after our stand-down period, we began working up the ship's readiness level by qualifying in a series of required training exercises. One of those exercises was becoming qualified in naval gunfire support. In those days, there was a dedicated island in the Hawaiian chain, Kahoolawae, designated for that purpose.

We had completed some of our qualifications during the evening hours and were scheduled to complete the remaining exercise for our qualifications the next morning. I was entrusted for the first time as an unrestricted underway OOD on the midwatch (midnight to four a.m.). The captain's written orders for me that night directed that we remain a certain distance from the island so we would be in a position at first light to resume the gunnery exercise.

It was a typically clear night when I noticed what appeared to be lightning to the west of us. A few minutes later, an ominous atomic mushroom cloud appeared where the lightning had been moments before. I awakened and advised the Commanding Officer in no uncertain terms of the incident and immediately called away the ship to general quarters (battle stations).

A record was set in manning battle stations since the mushroom cloud appeared to be in the general bearing of Pearl Harbor. We all apprehensively wondered if Pearl Harbor had again been attacked. We checked the fleet broadcast originating at Pearl Harbor and found they were still in business.

It would later be determined that the mushroom cloud was the result of a high-altitude A-bomb test shot that had been conducted near Johnston Island several hundred miles away from us. There were some anxious moments as we sorted out this information. What a way to qualify on my First Officer of the Deck night watch underway. Wow!

A Rescue at Sea in the Indian Ocean Saved Fifty-Six Survivors

Years later in 1973, when I was the Commanding Officer of a destroyer assigned to the Middle East Force—having recently relieved Classmate Bill Peerenboom's ship in the area—we were operating independently and proceeding from Djibouti on the Red Sea to our next port of call, Bahrain in the Persian Gulf.

During the early evening hours on the first day of our transit, we picked up a distress call from an Indian merchant ship some 150 miles from our location which was foundering and needed assistance. Maritime traffic abounds in this geographical area, and initially, we felt that other ships would surely come to the assistance of this distressed ship long before we could arrive in the vicinity. Still, some inner voice told me to proceed with all haste to the area.

Arriving on the scene in a little over four hours with heavy seas running, it was rough! There was no sign of a ship in distress. We started a search downwind of the last reported location of the ship with extra lookouts posted. Soon after the commencement of our initial search leg, a lookout reported hearing a voice from the water and threw a lighted float box in the vicinity because it was very dark.

As we turned the ship around, another lookout heard the voice, and this time we saw the first survivor and, very quickly, we had him on board. Later, joining in the search for survivors were an Israeli merchant ship, a Japanese fish factory ship, and a Liberian tanker. In two days, as the on-scene commander of the search and rescue operation, we rescued fifty-six survivors from the stricken ship and received worldwide recognition for our efforts. A great feeling for our crew to save those men.

POSSUM IN THE CHURCH
Jim Paulk

THIS STORY SOUNDS LIKE one that might be told by a comedian on the stage of the Grand Ole Opry, but this is the real thing. When we moved to Southern California, it took us a couple of years to find a church home, but when our daughter started going to Luther League with a classmate, we decided to try out the local Lutheran Church. During our married life, Pat and I always attended Methodist or Presbyterian churches wherever we lived, but we liked the people there and joined Grace Lutheran Church in Huntington Beach. We became involved in the church and attended services regularly and, as per normal, sat in the same place on Sundays, about halfway down to the front on the aisle.

Our Pastor, Paul Johnsen, was not only a close friend, but our oldest children married one another, thus our families were close. Paul's sermons were special. He had been President of the Lutheran Bible Institute, and he not only preached from the Bible but did it with an engaging delivery.

So, one Sunday, he was well into his sermon, and the congregation was thoroughly engaged. Then, most of the congregation in front of us were suddenly kind of crouched up, not quite standing, looking down front. Pastor Paul kept preaching and didn't notice anything, but after a couple of minutes, everyone sat back down. This situation happened several times until it became obvious there was a distraction on the floor in front of the pulpit. When Pastor Paul paused his sermon, we stood up and could see there was a possum (Us rednecks never use opossum, though we know what the critter is and the proper spelling, we are in Georgia, not Australia and culturally incorrect) on the floor in front of the pulpit.

Eventually, after Pastor Paul stopped his sermon, he asked the ushers, "Come down front and get this thing." By now, the congregation could see that a possum was darting in and out of an opening in the bottom of the chancel area. There was complete silence in the church, the congregation was paying close attention to this unfolding circus. Pastor Paul asked again for the ushers to come down front to remove this animal. Folks turned around to watch the ushers come forward with a trash can or something, but they didn't make a move.

I wasn't moving either, but Pat elbowed me in the ribs, and said, "Go down there and get that thing." So, I got up and hardly noticed that everyone was watching as I walked down front. I was focused on that wild animal with sharp teeth and claws that could rip me up pretty good. Now, this possum had to be nervous and afraid and quite obviously, had not been paying attention to the sermon. It was one frightened animal, and I had no trouble with him thankfully.

Wiggling my left hand to distract him, I picked him up by the scuff of his neck with my right hand and walked back down the aisle with him and toward the front door with all eyes on me. At this moment, I had the full attention of the pastor and the congregation—they were watching me intently. Maybe they were afraid I might let it go in the church again or drop it accidentally, but I went out the door with it and, after walking some distance away and advising Mr. Possum to stay out of our church, I let him go. He scampered away, free at last!

After walking back into the church and taking my seat next to Pat, she whispered, "Pastor Paul was talking about you the whole time you were gone. He told the congregation you were the best fisherman in Huntington Beach and an outdoorsman who grew up in Georgia." I knew that was not completely true because I knew many who were better fishermen but appreciated the compliment anyway. I was thankful the possum cooperated—he could have ripped me up good!

The lesson? Go to church, no telling what you might see.

BOB NEVIN, HYDRONAUT
Jim Paulk

WHEN WE MOVED to Georgia after thirty-seven years in Southern California, the week we got here, Bob Nevin, a 1957 classmate and good friend, called to welcome us to Kingsland and offered his help.

"By the way," he said. "Can we pick you up Friday night for the football game?"

"Are you talking about high school football? I haven't been to a high school football game in fifty years," I told him.

"Yes, it's the biggest thing in Camden County," he said.

From that point on, five of us went to every

local game and occasionally an out-of-town one—no wives, a man's thing. The Camden County Wildcats had won several state championships and had a football field like a professional one with facilities to match. They started the kids off in middle school, running the Winged-T formation so by the time they got to high school, they were ready. Rain, cold, or red-hot weather, we were at the games. Bob also helped me with planting and harvesting our crops in two farming areas; we were two close classmates and enjoyed activities together.

Like me, I knew Bob had qualified in submarines, and in addition, had been the Executive Officer of our local Kings Bay Submarine Base and had retired here. He was originally from Dayton, Ohio, but grew up and graduated from high school in Grosse Isle, Michigan, before attending Annapolis.

At some point, I learned Bob had been a Hydronaut, which is even rarer than an Astronaut. After graduating from the Naval Academy, Bob went to the Destroyer USS *Holder* (DD-819) for a year, Submarine Officer's School, USS *Cobbler* (SS-344), and then to the University of Washington to obtain a degree in Oceanography, and after a tour in the USS *Spinax* (SS/SSR-489), his career changed. He was assigned to do deep-sea ocean study work and to take charge of the bathyscaphe *Trieste II* (DSV-1) with the title, Officer-in-Charge and Chief Pilot.

Initially, his ocean study was off the coast of San Diego, near the Navy-owned, San Clemente Island about fifty miles offshore where he trained his crew of thirty-five officers and enlisted men, it was quite an undertaking to get the bathyscaphe to the dive location and prepared for the diving operation. Bob was not the only pilot of the craft and though it had some maneuverability, diving and surfacing required special maintenance issues. For buoyancy (to go up) purposes, the submarine-shaped, sixty-seven-foot-long float was filled with twenty-eight thousand gallons of volatile aviation gasoline which was lighter than water, and for diving, the float hopper was loaded with tons of #10 iron ingots to make it heavy enough to go down. For surfacing, iron

ingots would be released from the hopper, and the buoyancy of the aviation gasoline would cause them to rise to the surface. Hopefully, every time!

To prepare for diving operations and since the aviation gas volitivity made it very hazardous, the crew used their Boston Whaler boat to handle the long hose for filling the float some distance away from the ship providing the gasoline. A designated "sniffer" in the crew was responsible for seeing that no dangerous gas fumes were near the float during any operation and especially the task of loading the required tons of iron ingots into the hopper container—a laborious task for the crew as well. The preparations for the dives and maintenance took days to do so safely. The pressure-resistant sphere containing the crew was on the underside of the float thirteen feet aft of the bow, and the crew boarded it from a hatch on top of the float and a ladder to a water-tight hatch on the sphere.

When the *Scorpion* (SSN-589) went down on May 22, 1968, the decision was made to send the *Trieste II* to the area four hundred miles southwest of the Azores to examine the debris field. The general area where the boat went down with a crew of ninety-nine officers and enlisted men was located by the sound of the boat's tanks imploding with the sound waves measured from different sites to triangulate the location. Finally, the exact location of the *Scorpion* was found, and all was ready for Bob, his team, and the *Trieste II* to make their dives on the boat.

Before the *Trieste* crew left San Diego for the long journey, Bob assumed another responsibility, he became the Officer-in-Charge of the *White Sands* (ARD-20), an ocean-going drydock that would carry the *Trieste II* in its "well-deck" and use their cranes to lift it in and out of the water. With a complement of 112, the 292-foot-long, 81-foot-wide drydock was a perfect platform for moving, maintaining, and lifting the *Trieste II* in and out of the water. Now all that was needed was a tow, and that was supplied by the USS *Apache* (ATF-67). Bob indicated the accommodations on the *White Sands* were the most comfortable ones he ever had while at sea, (he was living like an admiral!).

The eight-month journey began with the *Apache* towing *White Sands* with the *Trieste II* on board, and they headed south along the Baja California Peninsula for the western entrance to the Panama Canal. After passing through the Canal without incident, the tow rig headed north for Mayport, Florida, near Jacksonville, to refuel and resupply for the last leg of their trip to the spot four hundred miles southwest of the Azores Islands on the edge of the Sargasso Sea, where they would meet the USNS *Victoria* (T-AK128) for resupply and the USS *Ruchamkin* (APD-89) as the on-scene support ship.

With the *Scorpion* location pin-pointed by the oceanographic research ship USS *Mizar* (T-AGOR-1) on the bottom about eleven thousand feet below the surface, it was now up to the *Trieste II* to verify the wreckage as that of the *Scorpion* and to examine the debris field surrounding it. As it turned out, the first dive was the only dive where the bathyscaphe came down close to the *Scorpion*, and they came down exactly on top of it, just aft of the submarine's sail—an amazing accomplishment. On all other dives, they had to search for it, sometimes for hours, before locating it again.

There were nine dives by *Trieste II* between June 13, 1969, and the last dive on July 31, 1969, which provides one with an idea of the length of time required to prepare it between trips to the bottom. Trips lasted from eight and a half hours to about sixteen. For a pressure-resistant sphere area of 7.9 feet in diameter, three men were very cramped in such a small space seated between a lot of equipment. The small propellers on the *Trieste II* provided thrust, but the difficulty in operating it was in maintaining depth control, which they solved by deploying the anchor and adjusting the cable length to hold them at the desired height above the ocean floor. They plowed the whole area around the *Scorpion* with that anchor and a wider area when they had to search for hours to find it. The ocean floor also was littered with iron nuggets by the time the mission was over, a testament to their activities there.

One crew member sat behind the pilot and copilot to monitor a color TV to direct the pilot in their search and for filming and photographing the *Scorpion* and the debris field, the primary purpose of the venture. There was also a sonar and other sensor equipment available to the crew. The pilot and the copilot could see a small area through the six-inch thick acrylic glass "eye," but the investigation direction came from the person monitoring the external camera and the visual display on the TV screen where a larger view was obtained. Bob took this position on several of the dives.

On one dive piloted by Lieutenant Anthony Dunn, the *Trieste II* became stuck in the muck on the bottom and could not free itself. Releasing iron ingots did not help, they were firmly held to the bottom. The propulsion of the craft was not powerful enough to move the craft to free them either. What were they to do? This was a critical situation because there was no way anyone could rescue them. Finally, when nothing worked, they decided to try rocking the craft side to side by coordinating their efforts inside the crowded pressure-resistant sphere. After learning to move together, they were able to rock the sphere enough to free it. Fortunately, they came up with a solution to this dilemma or they would still be down there. Anthony was godfather to Bob Nevin's son, David, who often attended high school football games with us. Crew members grew very close during the perilous training and missions for the *Trieste II*.

Another noteworthy event occurred on July 21, 1969, *Trieste* was piloted by Lieutenant Saxon with Lieutenants Byrnes and Field aboard as they descended on their dive to the bottom of the Atlantic. On *Apollo 11*, Neil Armstrong, Buzz Aldrin, and Michael Collins were on the moon and Neil Armstrong made his first step statement. Think about the technological developments involved in these two simultaneous historic events. Like competitors everywhere, the *Trieste II* crews good-naturedly came up with, "The ocean's bottom is more interesting than the moon's behind."

On the final dive of the mission, Bob decided he would pick up a few things in the debris field using the hydraulically operated, manipulator arm, "just to prove we were here." One of the items was a sextant with the name *Scorpion* on it and other items that someday may contribute to the investigation into the cause of the *Scorpion's* demise though this was not their mission. On one dive, they thought a body had been found, but after a close-up

observation, it was identified as laundry. No bodies were found outside of the *Scorpion's* pressure hull.

With the operation over, the *White Sands* with the *Trieste II* on board was towed by the *Apache* back to Ballast Point in San Diego. All were safe, and the scientific data in the final report went to the Pentagon. The mission was completed, and the men of the *Apache*, *White Sands*, and *Trieste II* were decorated with a Navy Unit Commendation for their extraordinary efforts. Bravo Zulu!

A NAVIGATION NIGHTMARE IN THE GEORGIA STRAIGHTS
Bob Strange

MY SHIP, a Guided Missile Frigate, newly built in Seattle and commissioned in Bremerton, Washington, had been conducting tests in Dabob Bay as part of shakedown training. The US and Canada jointly operated an underwater test range there for evaluating torpedo performance. After our exercise in the late afternoon, the CO asked me (I was both the Navigator and the XO) if we should anchor overnight and transit the Georgia Straits during daylight hours. I told him I thought the straits were clearly marked and I was confident in transiting at night.

As we proceeded to our turn into the straits, darkness had fallen. I was obtaining accurate visual fixes every two minutes, and I made the recommendation to turn the ship into the straits in six minutes. The CO, who had been following progress and periodically checking the Surface Search Radar screen, asked me to verify the ship's position, which I did, and it confirmed our position and the time to turn. The CO then requested I observe what he was seeing on the radar screen.

When I looked at the radar, I saw a solid coastline on where I wanted to turn. Again, verifying my visual navigation position, I told the CO I was confident of our position and I recommended turning on time, which was so ordered. Both the CO and I envisioned if I was wrong, our careers would be over, and the Navy would have a seriously damaged brand-new ship attributable to an inexperienced crew.

We proceeded into the straits without incident although there was a lot of trepidation and apprehension on the bridge. As it turned out, the false radar return was caused by tide rips. After that, we had no further problem transiting the straits to Puget Sound, Whew!

BUMPING QUEEN MARY *FROM DRYDOCK*
Bob Strange

AS A NEWLY COMMISSIONED missile ship that had been delayed by almost a year due to a commercial shipyard strike, we were endeavoring to complete all the required shipboard evolutions in preparation for our first Western Pacific deployment with our squadron. In those days, the powers that be directed that whenever possible, each deploying task group should have at least one missile-equipped ship assigned to the group. As we were the only missile ship available whose schedule fit that category, it was decided to deploy us within six months of our commissioning.

One evening, as we were returning to San Diego to drop off observers, we lost steering control in the channel, and before regaining control, we struck one of the channel marker buoys, damaging both our rudder and our single propeller. Investigation revealed a material malfunction had caused the problem but it would be necessary to drydock the ship for repairs.

Queen Mary cruise liner, which had been acquired by Long

Beach to ultimately become a convention center/hotel at the waterfront, had been receiving lots of press on its progress and was scheduled to undergo an extensive availability in the Long Beach Naval Shipyard Drydock. All preparations had been made for drydocking the *Queen* the next day when it was discovered that due to operational commitments, we needed the drydock.

Much to the consternation of the public fans and the press, *Queen Mary* was delayed by a week for her scheduled drydocking. Repairs on us were readily accomplished and we made our scheduled deployment as the sole missile ship escorting the carrier *Ticonderoga* Task Group. Sorry, *Queen Mary* fans.

MARINE OFFICER ESCAPADES
Larry Ingels

Adventures in Indiana

BETWEEN 1962 AND 1965, I was on recruiting duty with the Marine Corps in Indianapolis. The following are some of the memorable events during those years.

We had an Administrative Sergeant who was a tough cookie, someone you didn't want to mess around with. Once in the Philippines on courier duty, he shot a man he thought was threatening to take his classified courier mail. From time to time, the Sergeant would escort an AWOL Marine, delivered by the MPs from nearby Ft. Benjamin Harrison, over to the local Armed Forces Examining Station (AFES) for a physical exam before he was locked up in the local county jail. There, the AWOL Marine would await orders and an escort back to his unit.

Then one day, our Sergeant escorted an AWOL man over to AFES for his physical. The medical facility was located on the second floor of the building. When he took the man's handcuffs off for the physical, the AWOL man decided to make a run for it over to the stairs and down. The Sergeant reacted immediately, took out his pistol, and fired a warning shot at the stairway over the man's head. You can imagine the sound of a .45 caliber pistol shot echoing through the building.

The Sergeant ran over to the stairs and down after the man. There was a head (Men's Room) at the bottom of the stairs. At the entrance to the head, the Sergeant yelled, "All right, come out of there you son of a b---h!" Well, the first person to respond was this poor draftee. He came out of the toilet stall with his hands in the air and his pants around his ankles. He heard the shot and the commanding voice and wasn't taking any chances. Of course, the AWOL came out later as well, and things went almost back to normal.

That was one very memorable day in recruiting and in the AFES. I'm sure that poor draftee had a good story to tell the rest of his life.

As the Assistant Recruiting Officer, I had the job of writing up any vehicle accidents involving our official vehicles and submitting them to our district headquarters in Kansas City. During this time, I recall only two unusual accidents. The first accident involved our Officer-in-Charge who was driving out of town to visit one of our substations in a nearby city. It seems that at a traffic light, there was a truckload of scrap metal next to him. As they both started up, a hot water tank from the top of the truck fell off and hit the top of our recruiting sedan. I wrote up the report about our vehicle involved in an accident with a hot water tank and submitted it to district headquarters. I didn't hear back from them.

A short time later one of our recruiters, Sergeant "Birdie Deadwiler," a six-foot, 180-pound Marine, was involved in a multi-car accident, The other two drivers—I kid you not—were Casey Jones and Mr. Wrech (pronounced "wreck"). I always felt our district legal office thought I made up these two investigations. Sometimes truth is stranger than fiction.

Another time as a Recruiting Officer, I interviewed most of the recruiting applicants and would later swear them in. If someone was considered "marginal" as an applicant, then two of us would interview the applicant and compare our observations. On one occasion, the Sergeant Major interviewed someone and then sent him to me for a second interview. After talking to the applicant, I decided he wasn't exactly a "ball of fire," but maybe we should take a chance on him. I then proceeded to swear him in. I said, "Raise your right hand and repeat after me." He raised his right hand. Then I said, "I, state your name."

The applicant repeated, "I, state your name."

I then said, "Put your hand down." At that point, I decided he didn't quite meet our standards, so I decided not to swear him in. However, he did follow my instructions *exactly*.

Desert One

In the mid-1980s, I was an Assistant Facilities Engineer in Quantico, Virginia. It was during this time that President Carter approved an attempt to rescue the hostages from our Embassy in Tehran. Apparently, during the staging for the rescue, two of the helicopters collided on the ground, causing the decision to scrub the attempt and return the rescue teams to the United States.

The Marine Corps men in the operation of the mission were secretly returned to Quantico, Virginia, and were housed at old Camp Upshur—a remote area of the base west of Highway I-95. They were purposely kept isolated so they could not disclose what had happened.

Fred Mount, another Assistant Facilities Engineer, and I got involved in the support operations at Quantico. One of the things we had to coordinate was to get the Marine Corps leaders of the now-aborted mission into the Post Exchange at Quantico when it was closed to get them some "free" civilian clothes so they could go up to Washington, DC, posing as civilians, and to debrief the command at Headquarters Marine Corps and the Pentagon as to what happened to the aborted mission. We organized the late-night shopping spree. I understand they picked out some nice clothes since the price was right!

We also had to collect the team's automatic weapons, which had the serial numbers removed, and to get the weapons returned to our base armory for further disposition. This we did.

I got a call from Headquarters Marine Corps and had one other specific mission given to me. One of the Iranian guides assigned to lead the rescue team was temporarily staying in the Bachelor Officer Quarters at Quantico. I was told his name was "Ali", and that I was to pick him up at a certain time and take him to the National Airport so he could fly back to his new home (part of the deal?) in Texas. I was told one other thing—be sure to collect the .38 caliber pistol from him. We didn't want him trying to get on a plane with a pistol. I did as I was instructed, picked Ali up, took him to DC, got the pistol from him, and dropped him at the terminal. Later, I turned the pistol into our armory. As far as I know, Ali got home safely.

I don't think the public or the press ever knew these men were at Quantico for the few days until they could get "demobilized." It was interesting to me to have that small part in ending the highly publicized mission failure.

NAMING A SUBMARINE
Bob Fox

AT ONE TIME, submarines were named for fish, but all that changed when politics reared its ugly head and names were no longer fish. Jimmy Carter was the President and a Democrat when the Electric Bost Company was to build the nuclear submarine, SSN-705. She was to be named *Chicago*, and the excited pre-commissioning crew ordered their crew items like baseball caps, coffee cups, blankets, etc. to commemorate her name.

Now, pay close attention, this becomes amusingly a little complicated. In 1980, President Reagan came to the White House with his Republican friends, and no new submarine was going to be named *Chicago* while they were in office!

The SSN-705 still was years away from commissioning and given a new name, *Corpus Christi*. So, once again, the pre-commissioning crew needed baseball caps and other items with the *Corpus Christi* name on them. The *Chicago* items became souvenirs and history.

Time moved on, and the Catholic Bishop in Corpus Christi, Texas, learned of the new name for the SSN-705. He strongly objected to naming a warship with the name "Body of Christ."

Corpus Christi and Texas overall being a strong Republican territory, the Bishop had political power, and the name was changed again. This time, the name was to *City of Corpus Christi*. So, in 1983, the *City of Corpus Christi* (SSN-705) was commissioned with new baseball caps for the crew to wear.

I always wondered what if the SSN-705 was named *Wahoo's Traditional*! And, by the way, in 1986, *Chicago* (SSN-721) was commissioned.

NATIONAL RECONNAISSANCE PROGRAM
Bob Rosenberg

———

I ROSE FROM A NONRATED officer (referred to as a Ground Pounder by pilots) to Colonel (thirteen years) in a fighter pilot-dominated Air Force and the first member of our Class to attain flag rank (twenty-one years), positioning me to shape and impact National Security Space support to Naval, Land, and Air Forces. Much of my career was in Joint Force assignments spent influencing National Security, National Space Policy, and the effectiveness of Land, Naval, and Air Operations through a space program focused on military operations and recognized by the Secretary of the Air Force as a role model of service, steadfastness, and valor.

Because my eyesight had significantly deteriorated during my four years at the Naval Academy, a commission as a Navy line officer was not available to me. To further serve my country, however, I turned to the Air Force Liaison Officer at the Academy to ask if the Air Force would accept me and thereby enable me to embark on a career with the potential to serve in positions of responsibility equivalent to the Navy line. The Air Force Liaison Officer responded in an almost prophetical manner, "Of course, son. I'm going to send you to guided missile school . . . Someday the Air Force will be in space."

In the early days of the Cold War, the world was a place of fear and uncertainty. Our founding fathers' hard-fought freedom of over 180 years was at risk, as the Soviet Union—the Evil Empire—sought to destroy our way of life. The threat was real, looming behind the Iron Curtain, and the need to understand what transpired on the other side was crucial for our nation's survival.

As members of the Class of 1957, we found ourselves serving in a highly classified intelligence organization, the National Reconnaissance Office (NRO). Among us were Phil Papaccio, Don Regenhardt, Johnny Sedano, Bud Coyle, John Disher, Ed Smathers, and me. Our journey in the race for our nation's survival began against a backdrop of a world in turmoil.

The quest for information behind the Iron Curtain was paramount. The Soviet Union's military arsenal, shrouded in secrecy, posed a significant threat to us. The bomber and missile gaps fueled concerns, and President Eisenhower's proposal for an Open Skies Treaty aimed to ease tensions. However, Khrushchev's denouncement left us with limited options.

The turning point came with the launch of *Sputnik* on October 4, 1957, shaking the world's confidence in the United States. The realization that the Soviets had the means to deliver nuclear weapons to our homeland heightened the stakes. Attempts to assess their capabilities through secret reconnaissance flights faced a setback after Gary Powers was shot down in his U-2 airplane in 1960, a shock to the world.

This series of events spurred the birth of National Security Space, a critical development that would alter the course of the Cold War. The NRO's early satellite programs shocked the world of reconnaissance. By 1960, the NRO had imaged 3.8 million square miles of denied area, surpassing the coverage provided by all previous U-2 missions. The strategic advantage over the Soviet

Union was regained, eliminating the need for sensitive aerial overflights.

The NRO's foray into space not only transformed the national security landscape but also showcased the potential of satellites to surpass traditional means. The ability to map Soviet air defense radars and detect ICBM launches turned the tide of the early Cold War. My classmates and I were fortunate to be part of this revolutionary era.

My involvement in National Security Space began in the late 1950s and early 1960s as a member of the Vandenberg test team. Assigned in 1958 to Vandenberg Air Force Base, California, I became a contributor to the initial development, test, and launch of *Agena* satellites *SAMOS*, *MIDAS*, and *CORONA*, vital predecessor programs of today's photoreconnaissance and missile-warning satellite systems.

Participating in system tests of the *Agena* satellites, I spent over three hundred days per year on temporary duty in Sunnyvale, California, at the Lockheed factory helping build, test, and then take satellites to Vandenberg AFB to launch photoreconnaissance and missile warning satellite systems. My role extended to modifying the Strategic Air Command *ATLAS* Radio Controlled Guidance Station at Vandenberg to guide reconnaissance payloads into orbit.

From 1964 to 1968, following graduation from the Air Force Institute of Technology with a Master of Science degree in aerospace engineering, I returned to the satellite business at the National Reconnaissance Office (NRO), satellite test center as mission controller for the *CORONA*, *GAMBIT*, and *SIGINT* satellites where I provided increases in target collection beyond the specifications. The transition from three-day missions to three to four weeks marked a period of intense effort to meet mission needs. The team and I responded to several requests from analysts in the intelligence community, using special manual methods to capture images of Soviet facilities as they requested.

A few of the early images are enclosed with this story.

TOMSK-7 SOVIET NUCLEAR TEST FACILITY, January 24, 1966.

Lop Nor Nuclear Test Facility Shot Tower PRC, December 8, 1965.

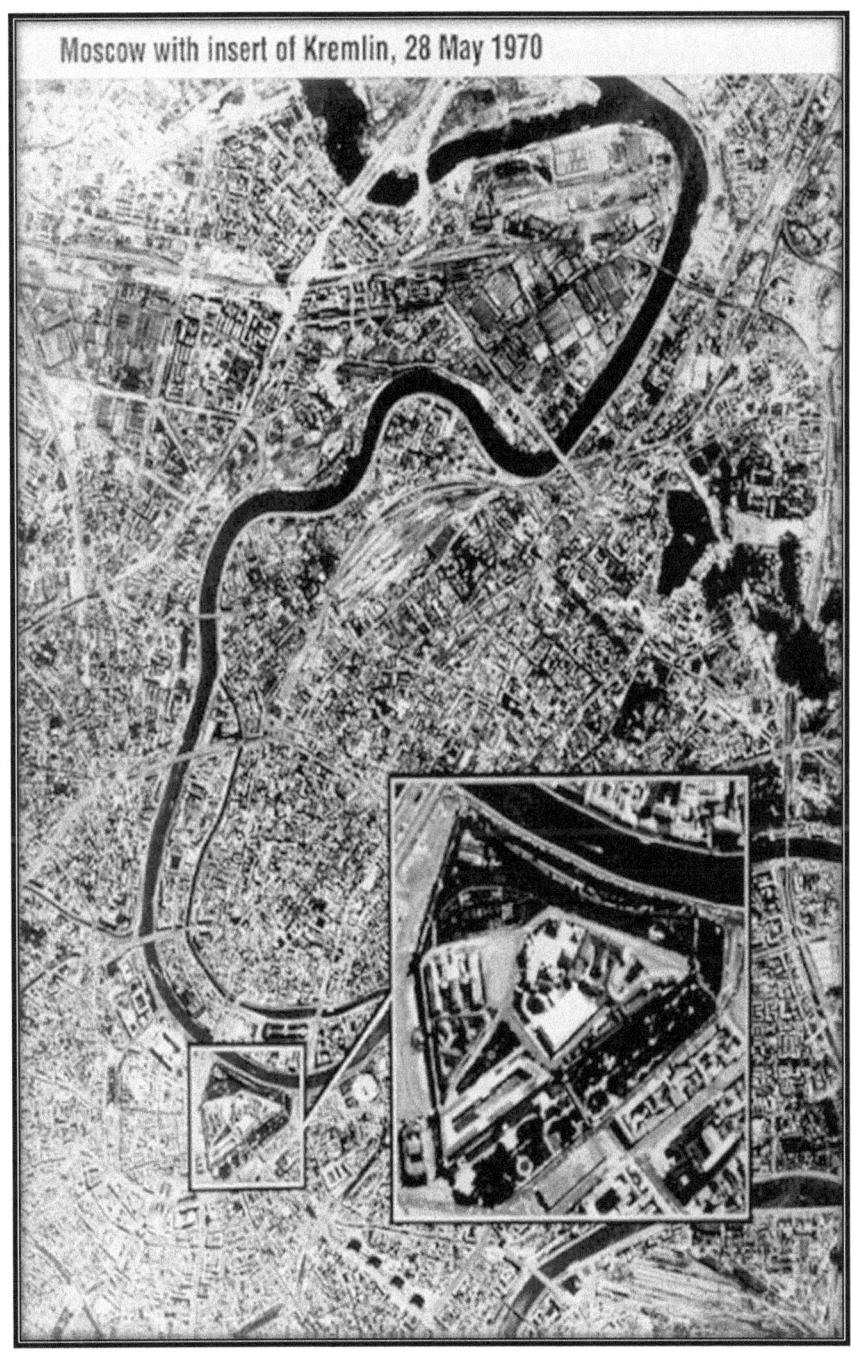
Moscow with insert of Kremlin, 28 May 1970

From 1969 to 1971 in the NRO, I was responsible for the acquisition of the C2 and mission planning and targeting software system for the last of the big film return reconnaissance systems. *HEXAGON*—that was a key collector for the Defense Mapping Agency (DMA) as well—made a significant contribution to the extremely accurate maps and targeting charts of today.

From 1973 to 1975, I was Deputy for Programs and then Acting Director of the NRO Staff serving in NRO Headquarters, Pentagon—the location where we directed and controlled programs supporting the military services and the intelligence community. We played a key role in providing systems that made it possible for the United States to see, hear, and know with certainty what was going on in the Soviet Union and the rest of the Communist world.

From then until the late 1970s, I had the awesome assignment on the National Security Council (NRC) staff at the White House, under the Assistant to the President for National Security Affairs in both the Ford and Carter administrations, where I led the efforts in policy formulation and decision-making for national security requirements, programs, and budgets related to intelligence, telecommunications, and US space programs; and was instrumental in establishing initial presidential funding for the Global Positioning System (GPS) to help my classmate Brad Parkinson's efforts to develop the system for military and civil use.

I, as a colonel, even won a shouting match with the Four-Star Director of Central Intelligence (DCI) when I told the President we needed to restore the major imagery satellite budget cuts he and OMB proposed. The DCI said, "Mr. President, Rosie doesn't know what he's talking about. I have more money than I need." I succeeded in winning that battle as well as getting the President to overrule and direct the DCI to acquire the system our warriors depend on still today for the night and all-weather support.

From 1983 to 1985, I was the Vice-Commander in Chief of NORAD and Assistant Vice-Commander of Air Force Space

Command, Colorado Springs, where I continued direct involvement in the operational use of National Security Space Systems. Finally, from 1985 to 1987, as the Director of the Defense Mapping Agency, I oversaw a $2.3 billion modernization program using the most modern imagery sources of the NRO.

Government/Civilian/Business /Community Career Highlights

While Executive Vice President and General Manager of Washington Operations, Science Applications International Corporations (SAIC), I continued an influential role for National Security, Space, and Intelligence communities by serving on National Security advisory boards for thirty years.

I led the way in advocating the value of Space for National Security. General John Raymond, Commander US Space Force, wrote, "You have been a hero to us all, and helped set the foundation for today's United States Space Force!"

I persuaded the National Security community to refocus NRO operations, giving the operations and requirements of the Land, Naval, and Air Force equal priority with national intelligence collection efforts. As a player in the DCI-directed Gates Blue Ribbon Committee on imagery after Desert Storm, I helped establish today's vital National Geospatial-Intelligence Agency.

As GPS Independent Review Team Chairman, I persuaded the Air Force to establish a strong focus on improved accuracy and availability for all military operations, thus advancing GPS as the international gold standard for space-based position, navigation, and timing.

These experiences, from the early days of National Security Space to my diverse roles in leadership positions, shaped not only my military and follow-on business careers but also contributed to the safeguarding of our nation during a critical time in history. The journey was challenging but the knowledge gained and the impact on national security made it all worthwhile.

PEARL HARBOR ATTACKED
Nancy Piper, Wife of Earl Piper

MY FATHER WAS A CAREER NAVAL officer and was commissioned from the US Naval Academy in June of 1934 and retired in 1960. My mother followed his ship from port to port (San Diego, Norfolk, Bremerton, Honolulu, Long Beach, Boston, etc.) living in houses, apartments, and hotels. As the family grew, we all became accustomed to our life of frequent travel, dockside goodbyes, and happy homecomings since Daddy's shipboard duties typically kept him at sea and away from home.

When Daddy's ship was assigned to the home port in Honolulu, TH (Territory of Hawaii), we followed him by passenger ship from the States and found a place to live in Pearl City, close to the main fleet anchorage. Such was the case in May of 1941. In Hawaii, Daddy was the gunnery officer on the USS *Benham* (DD-397) a single-stack destroyer. In Hawaii, he frequently deployed to sea but was often in port on weekends so we could enjoy family time with him more than usual. Although just four at the time, I do remember playing with my baby brother, and with our neighbor's two children next door.

Late in November of 1941, Daddy's ship left port as part of a task force of ships under the command of the legendary Admiral Halsey. Their mission was to transport several squadrons of fighter aircraft aboard two aircraft carriers to Wake Island to beef up US defenses there. At that time, there were many warnings of possible hostile action by the Japanese. When the Japanese attacked Pearl Harbor on December 7, 1941, Admiral Halsey's task force had not yet returned to port and its valuable aircraft carriers, thankfully, were out of harm's way. On that Sunday morning, I do remember hearing loud noises from the bomb explosions less than a mile away in the anchorage.

My mother bravely kept my brother and me away from the ensuing chaos and frantic activity outside. While taking laundry off the clothesline in our backyard, she could see the pilots in the cockpits of the Japanese planes as they swooped over the harbor. Our next-door neighbors were Christian missionaries to Hawaii; they came to our rescue. Their family name was Higuchi. Yes, they were Japanese. They drove us to the Pali Ridge automobile tunnel, which was being used as a temporary bomb shelter. We stayed there most of the day, through the second air attack, and returned home that night.

Although the Higuchi family was interrogated after the attack, they were permitted to stay on the island and were never evacuated to the holding camps in the States. Admiral Halsey's force returned to Pearl Harbor the day after the attack to a smoking, heart-wrenching scene. We saw Daddy only briefly and then he was off to sea again after the ships were refueled.

I have often thought about my "Navy life" growing up and how my mother, especially, rose so admirably to the difficult situations we faced. She always kept me feeling happy, safe, and secure. The Pearl Harbor experience is my once-in-a-lifetime story. My upbringing experience is a narrative of tough love and faithfulness.

SHIP STABILITY OR LACK THEREOF
Wilson Whitmire

IN 1974, I WAS completing the second year of what was supposed to be five years of duty in the same geographical area as specified for those having previously completed a command tour, a measure intended to reduce the cost of moving officers and household goods.

Much to my elation, I was rescued from Washington duty by classmate Charlie Noll, a submarine detailer at the time, who gave me orders to commission and command the USS *Point Loma* (T-EPF-15) in San Diego. *Point Loma* was to be assigned to the Pacific Submarine Force to support the deep diving, bathyscape *Trieste II* (DSV-1). Built as a Thomaston class landing ship dock and modified for arctic operations, her class of LSD was designed to float landing craft on and off, which required flooding down to a maximum of eight feet.

As *Trieste* required a minimum of twelve feet (putting the main deck aft awash), 1,492 tons of lead were added to the keel area. Following a nearly two-year conversion and modernization, sea trials were conducted off Long Beach, California. There were technicians and scientists with all kinds of instrumentation to monitor the ship as she flooded down.

Seas were moderate but after flooding down about nine feet, the ship started feeling very unstable with minimal righting movement. Both the XO (Jim Worthington) and I agreed it would be imprudent to flood down further, so much to the consternation of the many ship riders. I called off further testing.

As expected, many messages flew back and forth but Commander Submarines Pacific (VADM Charles Griffiths) backed the decision, and the David Taylor Model Basin (DTMB) in Maryland condescended to conduct further lab tests by building a thirty-five-foot model of *Point Loma*. The model when fully flooded capsized with the equivalent of four-foot seas. The lead DTMB designer came aboard the *Point Loma* and apologized profusely, saying they had somehow grossly miscalculated the original distribution of the lead. One must wonder what the fate of *Point Loma* might have been had scientific data prevailed over a couple of sailors' gut feelings!

Upon completion of a three-year command tour, I was reassigned as Chief Staff Officer of a nuclear attack submarine squadron in Charleston, South Carolina. While there, *Point Loma* and I were to again cross paths during her two cruises to the Atlantic.

The first revisit was to replace the CO, who had become ill and transferred ashore. *Trieste* was to make dives on the Cayman Trench, the deepest part of the Atlantic. This dive was partially sponsored by *National Geographic* and had hired consultant and lead scientist Bob Ballard, of *Titanic*, locating and filming fame. Dr. Ballard was one of three *Trieste* crew members when the bathysphere struck a cliff at nineteen-thousand-foot depth.

Talking to the crew on the underwater telephone, they thought they had ruptured the tank containing the positive ballast (highly volatile aviation gasoline) and that the craft was doomed. After releasing the negative ballast (about six tons of steel shot), the craft began a normal (six-hour) ascent, and the crew ascertained that what they thought was gasoline escaping was oil from the hydraulic reservoir for the craft's articulated arm. Although it was disappointing to abort the exploration of the Cayman Trench, I did have two-plus weeks of interesting conversations with Bob Ballard.

The second trip *Point Loma* made to the Atlantic was for *Trieste* diving on the sites of the *Thresher* and *Scorpion* losses, to be followed by *Trieste* picking up "something" in the Bahamas the Air Force had dropped. I went aboard *Point Loma* as the Officer in Tactical Command for the recovery portion of the Atlantic trip. This was without incident.

ST. MARY'S FOOTBALL COACH
Paul Roush

WHILE A MARINE CORPS CAPTAIN in my final year stationed with the Army at Fort Sill, Oklahoma as an artillery gunnery instructor, I was the (protestant) head football coach of the St. Mary's "Golden Knights" High School football team in Lawton. We practiced on the median strip of a divided highway in town after my working hours in the gunnery department at Fort Sill.

Since St. Mary's had no stadium, we played our home games at the field of the Fort Sill Indian School. The team had only won a single game in its previous five-year interscholastic history, but we won seven times the year I coached. Reverend Hayden, the head of the school, wanted to write to the Commandant of the Marine Corps asking that my tour at Fort Sill be extended for a year rather than have me execute my orders to an unaccompanied tour to Okinawa.

I told him if he wrote that letter, I would have to set the St. Mary's school on fire. He demurred, and I headed off to the Far East.

SUBMARINES CAN BE TARGETS TOO
Pete Baker

R ECENTLY, THE FORMER BOY soprano classmate from San Antonio, Al Hemphill, made a snide remark about surface ships being nothing more than "targets" for submarines. Let me tell y'all that the opposite is the case (and it can get you thrown out of bars).

After commissioning the Guided Missile Destroyer *King* (DLG-10) in Bremerton, Washington, in 1961, we proceeded to San Diego for shakedown training and firing of our anti-submarine torpedoes among other things. On a Friday, we teamed up with an old diesel boat for some anti-submarine warfare games and live firing of the torpedoes.

After several hours of "games," we fired the torpedo. Our sonar gang evaluated a hit and passed the information down to the submarine via underwater phone. The submarine reported back it was a clean miss and they were surfacing and then proceeding to San Diego for Broadway Pier and would see us there as we were both assigned "visit ships" for the weekend.

As the poor submarine surfaced about ten thousand yards off our port side, what do think we saw? Nothing but the rear end of a torpedo stuck in the submarine's sail. We couldn't break out the cameras fast enough. Nothing came from the submarine as she went to full speed and high-tailed it for home. We arrived about two hours later as we had one more exercise to complete. After mooring, we went over to look. No torpedo: and in its place was a patch of freshly painted canvas.

Ten days later we were in Pearl Harbor (as part of our shakedown cruise). A group of us decided to pay a visit to the Submarine Base Officer's Club with some pictures in hand. After a few cool ones, we started to hang our artwork with cool things written over them—such as "Submarirnes are nothing but targets for destroyers." We couldn't believe the lack of humor displayed by the manager when he asked us to take our pictures (or words, to that effect) and leave. Perhaps some of you may have seen the picture as it appeared in some publications over the next several years. More humor for some than others.

Yes, the Torpedo hit the submarine.

TALES FROM A FISH, CHICKEN, AND CORN FARMER
Don Beatty

Bangús Fish in a Philippine Market *(By Obsidian Soul)*.

A S I WAS GETTING CLOSER to the end of my Naval career, I was stationed in the Philippines and began to think about what I wanted to do after the Navy. All my experience and education had been oriented toward large organizations, but I had the idea I wanted to work for myself and outdoors but didn't know how to get there.

I had neither experience nor heritage in farming and at forty-seven years old, I wasn't a good candidate for carving trails with the Forest Service. I somehow thought I might find something in the Philippines and had taken a different approach to that country than some others around me. I joined the Local Kiwanis club in Olongapo and became friends with the local officials and business people. I even started to learn the Tagalog dialect.

When a job in Manila opened at the Philippine American Mutual Defense Board, I applied for it, was selected, and moved the family from Subic Bay to Manila. We lived in a civilian neighborhood with all Philippine neighbors, and I became friends with several of them and was invited to go to many of their activities.

A few of them were getting started in what was then a fledgling industry, the "ranching" of fish. Before that time, their favorite fish (bangús) was raised only in earthen ponds. They were going to raise them in huge pens constructed of bamboo and fishnet, and there was a huge market for that fish.

I was often invited to go along with them on their business activities on weekends and evenings. As I watched their operation for more than a year and observed their problems such as typhoons, securing nets underwater, and security—I helped them wherever I could with weather reports. The Philippines had no good civilian weather forecasting, engineering for their pens, or underwater repairs.

I used my scuba experience to build *hookah* rigs (air compressor and hose rigs for divers) and instructed them in their use. I also helped with their communications for security.

One day, a group of them arrived at my house and asked if I would like to join them as an independent operator doing the same type of business. They introduced me to a man who told me he would personally teach me the business from production to marketing. I jumped at the chance, found my wife in one of her weak moments to get her concurrence, retired from the Navy, sold our home in the United States to raise cash, and became a fish farmer.

We would plant fingerlings in a large area with fine netting where they would eat the natural food in the salt water. As they grew, we would change the size of the netting. At harvest, we caught them in a net like a purse seine that is used in tuna fishing at sea, dumped them into a boat filled with ice water where they died instantly, and hauled them to market where they were sold by the washtub full. On a given night, we would sell one to two hundred tubs of fish.

The business was terrifically exciting and enjoyable, but there was one big problem, the paychecks came only once a year! I had a large native crew who counted on me for their livelihood, and if there was a poor crop of fish to market, it was a long time between paychecks. I began to look for a business with a more frequent cash flow and decided to become a poultry farmer in addition to my fish operation.

I approached a large corporation that sold chickens and asked them to give me a contract to grow chickens. After their initial surprise, they agreed and gave me advice as to where to locate the farm if I could find land to lease. I needed ten acres of land for the farm and found a parcel that was in the area they suggested, but it was one hundred acres in size. The owner would not break it up, so I agreed to lease the hundred acres.

After reading a huge book from cover to cover on everything from poultry house construction to chicken husbandry, I built and began to operate a poultry-raising business that reached five hundred thousand chickens per year. Soon a decision had to be made as to how to utilize the remaining ninety acres of land that had been historically a sugarcane farm. I then decided chickens could always eat corn and started a corn farming operation.

However, there was a problem with farming in the Philippines as there were only two seasons—wet and dry. I had to find irrigation for use during the dry season to efficiently grow corn, and there was none available in the Philippines.

That's when I became acquainted with some Israelis and had a sprinkler system shipped in from Israel. The first day it was put into operation, people came from miles around to view it. The

Israelis decided I would make a good guinea pig for trying to introduce new vegetables to the Philippine market and with their help, I began to grow a bit of everything from melons to eggplant. I became the largest producer of "baby "corn in the islands and introduced "super sweet" corn to the Philippine market.

Everything went well except for the typhoons, which destroyed my entire operation twice. Trying to get started again in a foreign country with bank interest rates as high as 42 percent was a huge challenge.

My friends from the fish business decided to try to grow shrimp in the same way we grew fish. I was selected to be the main project monitor for the group and found myself living on a gorgeous beach under palm trees in a one-room hut with a thatched roof that had been specially constructed for me and which had the only flush toilet within twenty miles. No TV, only short-wave radio and the nearest telephone on top of a mountain that took two hours to climb.

When my wife, Kay, would fly from Manila to visit me, the local people would bring their children to see the "white woman" and to see my toilet. After a year, I saw the shrimp project wasn't going to work and returned to Manila.

The business environment, the natural environment, and the political situation had by that time deteriorated to the point that I decided to return to the US and start another career. After having spent fourteen years in the Philippines, I felt sad about leaving. It had been a very exciting, challenging, and sometimes dangerous period in my life I still often reflect upon. I had survived rebellions, insurgencies, and business problems most Americans would not believe, but was left with a treasure chest of great memories.

THE FAREWELL BALL DRESS
Larry Magner

LOUISE MAGNER AND JOAN ZIMMER were the best of friends and, although their husbands, Larry (me) and Emory, knew of each other at the Academy, it wasn't until their diesel boat time together and deep draft (surface ship) days that they and their families became close friends.

Emory Jr. (Zimmie) was about to graduate from Annapolis when Joan put in a desperate call to Louise and said, "Can you believe it? Emory has just broken up with his fiancée, and it is only three days until the Farewell Ball! He's devastated."

Louise replied, "That's terrible. What is he going to do for a date?"

"I was hoping Carolyn, (our daughter) might consider going with him. He always did have a crush on her."

"Carolyn's not in town, but Laura (another daughter) is home from college and might be interested."

Joan said, "Let me check with Zimmie, and I'll get back to you."

Joan called back an hour later and said, "He'd be delighted to take Laura, even though he hasn't seen her since she was in high school."

After much cajoling, Laura finally agreed to go to the ball and said, "Okay, but it's going to cost you a new dress."

Louise replied, "Laura, dear, why don't you take a look at the dress I wore to the Farewell Ball twenty-four years ago?"

Louise went up to the attic and brought down a sealed box, opened it, laid the dress out on the bed, and cried out, "Laura, come on up and tell me what you think."

Laura came into the bedroom, took one quick look at the dress, and said, "It's beautiful, Mom, but it's a little out of style these days. I'd really enjoy having a new dress."

The next day, mother and daughter spent many hours looking at formal dresses in several upscale stores but to no avail. Laura said, "We'll give it another go tomorrow, Mom, but now I have to go to Joan's house to meet Zimmie and get more details on the Farewell Ball."

Upon arriving at the house, she was greeted by Joan who said, "Zimmie will be here in thirty minutes. By the way, have you thought about what to wear to the ball?"

Laura replied, "Funny you should ask. Mom and I spent the whole day looking, but I couldn't find anything."

Then Joan said, "Look, Laura, I just happen to have the dress I wore to the Farewell Ball in 1957. Would you like to see it? If it fits you and you like it your problem is solved."

They went upstairs and Joan laid the dress out on the bed. Laura took a quick look at it and immediately said, "Thanks for offering it but I'm afraid that dress just won't work for me."

Several hours later, Laura stormed into the house and yelled, "Mooommm! What are you trying to do to me? I told you I wanted a new dress. Why did you give Joan your dress so she could try to get me to wear it?"

"What are you talking about? My dress is still laid out on the bed upstairs in case you change your mind."

Laura, somewhat puzzled, said, "Did you know Joan when you went to the ball?

"No. We met many years later."

Laura laughed, "It's a good thing you didn't bump into her at the ball because you both had the same dress on that night."

"It's impossible. I got mine in New York City. She's from Cleveland!"

Laura gave up on a new dress and went to the ball wearing her mom's dress. Zimmie had a great time and so did Laura. Louise and Joan laughed and agreed that it was a darned good thing they didn't run into each other at the Farewell Ball.

Laura told her mom afterward about having dinner at a nice restaurant before the ball. She said, "You know, Mom, when I walked into that restaurant, everyone stared at me, so I just pretended I was Scarlet O'Hara."

True story, this ain't no lie!

WORMS
Jim Paulk

WHEN I WAS A YOUNG LAD, one of my great joys in life was going with my paternal grandfather to his farm in Ocilla, Georgia, located about one hundred miles west of Brunswick, where I lived on the Georgia coast. Back in the 1930s, adults didn't pay much attention or regulate their kids' lives very much. So, going to Grandpa's farm was always an adventure to see and do things I couldn't do in our small town—without any adult supervision either.

Now, I was born ten miles out of town, but we moved to town when I was four so I could go to school because buses didn't run that far out of town. Our first Paulk in America came in 1692 from Scotland, and though he settled in Concord, Massachusetts, subsequent generations migrated down to South Georgia where farming was better. I got to move cows from one pasture to another with Grandpa's farm manager, ride in his pickup truck, and help with chores. When they harvested the tobacco, I rode on the sled with the leaves being pulled by a mule to the barn where the tobacco was cured before taking it to the auction—exciting things for a little boy.

But, but, but, my real plan—always my plan—was fishing! Grandpa had a 125-acre millpond with fish in it, I didn't care how big they were. At one time, the mill wheel was turned by the water running over the dam, and kernels of corn were placed in a round stone wheel cut out so a large stone roller went around inside of it to grind the corn into grist or cornmeal. I never saw it in operation, but it was a fascinating piece of equipment and still there for me to see.

In this millhouse, lying on the floor was a collection of eight-foot bamboo poles, to which seven feet of linen line was attached with a large hook on the end, and a wine bottle cork was used as a float. Taking some of these rigs out in the flat-bottomed rowboat was Heaven for this little boy. I wasn't strong enough to paddle (no rowing, couldn't do that) very far, but I could go far enough to find some fish if I had bait—an important detail.

In one of my favorite movies, *A River Runs Through It*, two brothers begrudgingly await another person to show up for a day of flyfishing, which is done using artificial baits like feathers tied around a hook. They commented that the other person running late would probably show up with a coffee can of worms, which he did to the chagrin of the brothers.

Now, I needed worms. I wasn't flyfishing, and my grandpa had told me how to get worms—pay attention, this is a lesson. He gave me a board about three feet long, two by one inch, with one end cut to a point. He told me where to go in the woods and gave me a red brick to use in this worm-finding operation. I was told to pound this stake into the ground with the brick, kind of deep, and then rub the brick across the top of the stake—keep rubbing until you see worms.

So, there I am, out in the woods, by myself, rubbing a stake with that red brick with nothing happening. With Grandpa being somewhat of a prankster, I was beginning to think he was pulling my leg. I'm there thinking, *This is kind of stupid*, but I kept rubbing that stake all the while thinking people were going to show up laughing at me. But suddenly, and magically, worms seemed to appear everywhere around me. This was crazy, but I picked up enough to put in my coffee can for a day of fishing in the pond. I would keep one pole to fish with for bluegill, and the others, I'd bait and toss in the water to retrieve later, they floated well—sometimes, I would find a good-eating speckled catfish on the line. This was my secret method of fishing, those poles didn't move too far, and I could always find them. Yum yum!

I'm told, in Alabama, this is called strumming for worms, but with another plank rubbed across the top, but in Georgia, the brick worked great. Choose your state method . . . kidding a little bit here! The football rivalry between Georgia and Alabama is legendary. Worm finding too!

Why did this work? The sound radiating in the ground seemed like rain to the worms, and they came to the surface to avoid drowning. It worked. Grandpa Paulk was a wise man and a person I truly admired. Yes, I caught a few fish (mostly bluegill) we cleaned and fried for dinner. So, go get you some worms!

UNDERGROUND WAR STORY
Roy Dahnke

EARLY IN THE DEPLOYMENT of the Minuteman Missile Systems in Montana, I was on assignment as a Combat Crew Commander in the Alpha Command Center. We handled all the communications regarding the status of fifty Intercontinental Ballistic Missiles with Strategic Air Command Headquarters, Home Base Headquarters, five missile sites, and any air or ground traffic moving through our area. The command centers were underground in hardened sites, with motion detectors and fences, as well as armed security guards with fully automatic rifles stationed above the underground site. Each missile command site housed only two people underground, the commander and deputy commander.

Because there was no such thing as an "inhibit override switch" to a launch command, other than within our underground command sites, all orders including launch commands came directly from SAC Headquarters. We, in turn, made the decision regarding the launch, and within ninety seconds, missiles could be launched to anywhere in the world. I won't comment on what kind of warheads they contained, but I'm sure you can guess. This was unique at the time, and the US Congress was very concerned. Therefore, we had many visitors from the US House and the US Senate touring the Alfa Command Site, which I often manned, and they asked lots of questions.

One day, I had a US Senator on board and was showing him around when he asked what we had in the file cabinets above our control panels. I explained that in addition to some food (my lunch), they contained cryptographic decoding books. He asked to see them. I replied that no one could see my decoding books other than me, and no one could see my deputy's decoding books other than my deputy. He responded that he was a "United States Senator" and would see them! I replied that even the United States President—who could initiate a missile launch—was not allowed to see them. He again replied that he was a "United States Senator" and *would* see them, and started to reach for the cabinet, which had no locking capability (mostly due to the requirement for rapid response).

As he reached for the cabinet, I drew my .38 revolver, placed it one inch from his chest cavity near his heart, and pulled the hammer back, with a very audible click. I stated that he might see the decoding book, but he would not walk out of the underground site. I then instructed my deputy to call the guards above ground to come down to the entrance to our underground capsule, advised the "United States Senator" that this tour was over, and escorted him to the eight-ton door, which I closed behind him as I turned him over to the security guards. I then went on with my normal business.

I never heard any more about this tour. As to what I would have done if he had continued to grab a decoding book, I leave it up to the reader to guess. The Senator made a good decision not to call my bluff.

WOMEN IN THE NAVY
Wilson Whitmire

(WHEN READING THE FOLLOWING, please keep in mind this took place in the early 1970s.)

Following my submarine command tour in 1972, I was assigned to the Officer Career Planning Board, a new office in the Bureau of Naval Personnel (BUPERS) intended to provide inputs to the Chief of Naval Operations and the Secretary of the Navy while bypassing the usual Washington Command layering.

Shortly after my arrival there, we were slightly reorganized with the arrival of Paul Keenan, who became Director of Professional Development. Paul had one assistant and a secretary, and the Career Planning Board—which I headed—came under his purview.

I had permanently assigned officers from each warfare community, including Special Warfare, and a Warrant officer. I also had the luxury of calling in other officers from the Washington area to participate in studies. We did three major studies (submarine, surface, and aviation), each lasting about six months.

We would routinely present preliminary recommendations to a board comprised of the Chief of Naval Personnel (VADM David Bagley) and the Pentagon Vice-Admirals heading the three major warfare communities. The final study recommendations were presented to the CNO/SECNAV and, as it turned out, all recommendations were accepted.

These included accessions, attrition rates, promotion points, command opportunities, the establishment of surface warfare schools in Newport and San Diego, surface warfare and surface command qualification requirements, approval and design of the surface warfare insignia, and the deletion of the requirements for aviators to be airborne for four hours a month to qualify for flight pay, instead qualifying them if they had sufficient time in operational squadrons before shore duty.

In short, many changes were affecting the warfare communities that came out of these studies. This was primarily because many study participants were almost all fresh from the fleet and had a pretty good perspective of what the operational forces needed. The CNO was enthusiastic about the results and upon completion of the last warfare study, directed the Officer Career Planning Board to do a study of women in the Navy to determine how they could be better utilized while achieving a more fulfilling career path.

I called in about fifteen officers, mostly Captains, representing a fair cross-section of personnel in the area. The senior woman officer in the Navy at that time was Captain Robin Quigley, attached to the Bureau of Personnel, who was a study attendee. The study took one day and all attendees, including Captain Quigley, agreed that women in the Navy were currently doing what they should be doing.

Previous study results had been presented in "dog and pony show" format in the Pentagon. This study had no recommendations other than maintaining the status quo. I drafted a letter containing my finding, which was given an up-check by Paul Keenan, signed by Vice Admiral Bagley, and forwarded to the

Chief of Naval Operations who, according to a witness, read it, uttered a few expletives, and trashed the letter.

He then requested that the Rear Admiral who headed the office—what we called the "touchy-feely" office—do another study. As a courtesy, the Admiral invited me to his study group meetings, and I did attend the first few where discussions concentrated on the placement of women on board ships and in airplanes and assignment to billets normally reserved for personnel going to assignments ashore.

There were also proponents of admitting women to the Naval Academy. I was a negative voice in these discussions with the Admiral often asking me, "Why can't we do this?"

I finally said, "Admiral, you can do or recommend whatever you want, but you know the attraction to the Naval Academy and a follow-on career in the Navy, for me, was the fact that it was a man's world—and quite frankly, I probably would have chosen a different career had it not been."

I didn't make much of an impression, as the CNO endorsed the study and many of you are familiar with the subsequent policy changes.

[*Compiler:* Today, women are not only at the Naval Academy and most branches of the Navy, but in leadership roles as well, including Superintendent of the Naval Academy.]

THE WORLD WAR I FOURTEEN-INCH NAVAL RAILWAY GUN
Bill Peerenboom

A S A PROUD GRADUATE of the US Naval Academy Class of 1957, I found myself nearing the end of my tenure as Deputy Director of Naval History, with a strong desire to leave a legacy. At the Navy Historical Center, I stumbled upon an old plan created by a predecessor Deputy Director—an intriguing idea involving a historic World War I, fourteen-inch Naval gun to be displayed at the Willard Park area of the Washington Navy Yard. This park, adjacent to the Navy Museum, featured various large-caliber Navy guns, but it lacked the impressive fourteen-inch gun.

The plan was detailed and comprehensive, outlining the existence of a complete "railway gun" located in Dahlgren, Virginia. It covered the logistics of transporting the gun by barge to the Navy Yard and assembling a short stretch of rail leading from the seawall to a concrete pad, where the gun would be mounted. Additionally, the plan meticulously accounted for all necessary expenses, including transportation, construction, renovation, and other miscellaneous costs. However, to bring this ambitious project to life, funding and approval were still required.

Determined to turn this vision into reality, I sought the support of my close friend and fellow Third Company Class of 1958 graduate, Jesse Hernandez, who held the position of Commandant of the Washington Navy Yard. While Jesse was enthusiastic about the idea, there was a potential obstacle—our Public Works Officer (PWO), known for being cautious and hesitant to approve ventures like this due to various regulations and constraints. He was affectionately known as "Doctor No."

Thankfully, I had previously connected with a young officer leading a reserve Seabee Company, whose members held their monthly meetings at the Navy Yard. They were eager for a challenging project. Recognizing an opportunity, I shared the plan with the Seabee officer, and to my delight, he and his team expressed enthusiastic support, believing they could complete the task in a single weekend.

Armed with this newfound backing, I organized a meeting with Jesse, the Commandant, and the PWO. I also invited the enthusiastic Seabee Lieutenant to join us. As we presented the plan, the PWO appeared puzzled by the young officer's presence, but he remained silent. Jesse voiced his approval, and when the PWO raised concerns and listed potential obstacles, I turned to the Seabee officer, who confidently assured us that his team could handle the project during their active-duty weekend. Jesse swiftly agreed, leaving the PWO with no room for further argument.

With the Navy Yard's acceptance of the plan, the last piece of the puzzle was securing funding. I approached VADM Joe Metcalf and passionately explained the project, requesting assistance in finding the necessary funds. True to his word, VADM Metcalf secured the funding, and the project was set in motion. I retired, but my successor, Chuck Smith '58, accomplished the job.

Today, the fourteen-inch naval gun proudly stands at the Washington Navy Yard, aimed at the Capitol, a testament to my initiative and determination. While I may not be around forever, my role in bringing this historical monument to the Navy Yard will be remembered and celebrated for many years to come.

Fourteen-Inch Naval Railway Gun in Willard, Park, Washington, Navy Yard.

FLAG LIEUTENANT FOLLIES (1)
Ted Kramer

ON MY THIRD SOLO DAY as Aide and Flag Lieutenant to the Commander of Amphibious Force Seventh Fleet, the Admiral and I were scheduled to fly to the Naval Air Facility in Atsugi, Japan, from our base at Subic Bay, Philippines, to attend a conference at the US Naval Base in Yokosuka, Japan. I had etched in my brain the fellow I relieved, telling me that the most embarrassing thing that happened to him and that almost got him relieved was leaving the Admiral's luggage on the plane, which took off with the Admiral's suitcase safely ensconced in the plane's luggage compartment destined for who knows where. So, on the entire trip, I repeated the mantra no less than a thousand times, "Don't leave the Admiral's luggage on the plane. Don't leave the Admiral's luggage on the plane."

When we landed at Atsugi, I quickly jumped out of the plane, grabbed the Admiral's luggage, deposited it in the trunk of the sedan the Naval Base had provided us, and started down the road to Yokosuka about twenty-five miles away, my chest swelling with pride that I had done my job with great aplomb. About halfway to Yokosuka, I was still congratulating myself when I started to think, *Now, did I get MY bag off the plane?* Oh my gosh! I did not, and it was halfway to the Marine Corps Air Station at Iwakuni, Japan, many miles away. I was scheduled to accompany the Admiral to a luncheon in civilian clothes the next day, but my clothes were about to become Marine Corps domain.

As soon as we arrived at Yokosuka, the Admiral jumped into a cab and headed into town and told me he would not be using his sedan so I could use it if I needed it. I made a few quick phone calls to Iwakuni and made arrangements for the Marines to capture my suitcase and return it to Atsugi on the next available flight and then on to Yokosuka by bus, but, obviously, not in time for the civilian luncheon the next day.

What to do! WHAT TO DO! Well, what I did was what any normal Naval Officer would do. I went to the Officer Club bar to let a scotch and soda help me.

Lo and behold, a solution was standing in the corner of the bar surrounded by a bevy of nurses—our good Fifth Company classmate and good friend, Ron Blackner, who was stationed aboard the USS *Constellation* (CV-64), tied up conveniently at a pier in Yokosuka.

Luckily, Ron was taking a break at the Officer's Club where fate managed to rendezvous me with him. I fought my way through the nurses and asked Ron for help and he said, "Sure, Ted, we'll grab a cab and head down to the pier, and I'll fix you right up." I told Ron the Admiral said we could use his sedan as it would be faster than waiting for a cab. So, we jumped into the Admiral's sedan and headed to the pier where the USS *Constellation* (CV-64) was tied up. As I parked the car next to the quarterdeck, I turned around to Ron and asked him what all the commotion was about on the quarterdeck—that is, bells ringing, whistles blowing, people running around, spy glasses trained on us, etc.

Ron examined the scene, uttered a short "uh oh," and asked me if I took the Admiral's two-star plates off the car.

I replied with complete Naval decorum, "Huh?"

They had also spied me with my uniform and aiguillette still on, and a tall man in civilian clothes sitting next to me. They assumed the worst. When they piped, "Admiral, United States Navy, arriving," and two half-inebriated junior officers stumbled up the gangway, I figured we had some big-time explaining to do. I told the Officer of the Deck that the Admiral had sent me down to the ship to pick up a package from his wife that Lieutenant Blackner carried to Yokosuka for him, and I thought the Admiral's driver had taken the plates off the car. I apologized for not checking. Lying is not a normal Flag Lieutenant trait but sometimes it comes in handy.

I thanked Ron for the clothes, and hurried past the quarterdeck as decorously fast as I could, saluted smartly, and departed the ship. I made it to the luncheon in Ron's snazzy Wyoming attire—less the cowboy boots and Stetson—but under the Admiral's questionable eye. I returned them the next day when my suitcase arrived from Iwakuma, and later returned to Subic Bay with both mine and the Admiral's suitcases well in hand.

[*Compiler:* Nice recovery, Ted.]

FLAG LIEUTENANT FOLLIES (2)
Ted Kramer

ON A CHILLY NOVEMBER DAY IN 1963, our flagship was anchored in the harbor at Kaohsiung, Taiwan, when the Taiwan authorities invited my Admiral, the Commander of Amphibious Forces, Seventh Fleet, to fly out to the offshore island of Quemoy to tour the fortifications and other facilities there. The Admiral invited me, as his Flag Lieutenant, to tag along, and I was delighted to participate in a new adventure.

We boarded a twin-engine propeller-driven plane and flew close to the water to fly under the mainland Chinese radar (only a few miles away from the island). When we landed, we were met by a host of Taiwanese military officials and given a most interesting tour. All the facilities—barracks, hospitals, gun emplacements, munitions depots, hangars, and everything except the small airstrip—were underground in fortified, thick concrete. It was most impressive. Each facility was connected by fortified tunnels accessed by locked steel doors. I don't think Fort Knox could have been more secure.

Lunch, also, was served in a fortified underground officers' mess which, like everything else, was impregnable and impressive. About halfway through lunch, the Taiwanese General stood up, tinkled his water glass with his spoon to get everyone's attention, and said he had an important announcement to make.

"We just received word from Subic Bay for our honored guests, Admiral Lee and his Flag Lieutenant. Lieutenant Kramer's wife has just given birth to a baby girl at the Subic Bay Naval Hospital and both mother and daughter are healthy and doing fine. So, gentlemen, let's all raise our glasses and give Lieutenant Kramer a hearty toast for his new daughter." Everyone in the mess stood, faced me, and toasted me. I was both a little embarrassed but happy at the same time. It was an extremely nice gesture and several of the officers came over to congratulate me after lunch was over.

We finished our tour and headed back to Kaohsiung and the ship and this Flag Lieutenant was extremely happy but at the same time, eager to return to Subic Bay to see my new daughter.

As I was unpacking my briefcase, the staff Communications Officer burst into my cabin and said, "Ted, have you heard the news?"

I said, "I certainly have, Commander Miller. It's the best news I've heard in ages! I'm so happy. I couldn't have received better news."

Commander Miller looked at me with somewhat of a disdainful look and said, "President Kennedy has been assassinated!"

Crash! Thud!

FUELING IN BERGEN, NORWAY
Mike Giambatistta

FOR INDEPENDENT OPERATIONS OF A SUBMARINE, refueling was an issue that had to be provided for at each port of call because we didn't refuel at sea from another ship as Naval ships could do. It was a snap in places like *the English Submarine Base* in Gosport, England, but because of our relatively high-speed surface transit between survey points (every sixty miles) in our current exercise, our boat, *Archerfish* (SS-311), with a fuel capacity of 110,000-gallons was often stretched to prudent limits between port visits. After fueling 24,500 gallons in Hammerfest, Norway, as the Engineering Officer, I began to develop reasonable confidence in the administrative system supporting *Archerfish*. The US Naval Attaché's office in each of the countries we visited seemed to work well with our supporting submarine force administrators back in New London, our Home Port.

It was the Fourth of July when *we* arrived in Bergen, Norway, after a couple of weeks at sea. Despite being eager to get ashore, I wanted to leave no stone unturned regarding refueling (my "Sea

Daddy," Miles Graham, always supplied sufficient "heat" to make sure I was not asleep at the switch). Consequently, I took the precaution of confirming our fueling arrangements for Bergen via radio while in transit to Bergen with the US Naval Attaché's office in Oslo. I was instructed to contact the ESSO (now Exxon Mobil) bunker depot in Bergen where all the appropriate arrangements for fueling had been made.

When we arrived in Bergen, we started the refueling process, and when it proceeded with the anticipated ease, I went below to get a cup of coffee, shine my shoes, and brush my best blues uniform (after all, it was July!) in keen preparation for a day of viewing the Bergen fish market and riding the cable car. My reverie was soon interrupted by a call from the topside watch reporting that some incoherent (to him) Norwegian wanted to see me about the 90,500 gallons of diesel that *Archerfish* had taken aboard. *What could this be*, I thought?

I put down the unfinished cup of coffee and went up the forward torpedo room hatch to find out what the commotion was all about. I produced all the documentation associated with the event (after all, it had worked up in Hammerfest a few weeks before). The gentleman's limited English and a total absence of even a single repeatable word of Norwegian in my vocabulary inhibited the ensuing exchange. Finally, through some gesturing and much intuition, it dawned on me that the ESSO guy was demanding payment.

In desperation (although the sun never set), I was getting edgy about the late hour for my departure ashore. So, I pulled out my wallet and produced my ESSO credit card. Much to my combined astonishment and relief, a huge smile spread over "Stig's" face as he snatched the card from my hand, copied all the data onto the form on his clipboard and requested my signature. What the Hell, I figured that before any charge got to me, the whole thing would be straightened out, so I signed!

This story was retold, with some embellishment, over the years in wardrooms and bars as part of the *Archerfish's* legend. Never did I hear from ESSO, although I never tried to get another card from them either!

HELICOPTER DOWN
O. C. Baker

THIS IS AN ACCOUNT OF ONE of my incidents while flying Marine CH-46 helicopters in the Purple Foxes detachment in Vietnam.

Corporal King was my .50 caliber machine gunner the day we were shot down near An Hoa. The rest of us left the helicopter and were lying down for protection in the rice paddy, but Corporal King stayed exposed in the helicopter with his machine gun and was firing these impossibly long bursts that were sure to burn up the barrel. I finally got up and ran back to tell him to fire shorter bursts.

What I found was that the firing mechanism on his machine gun was sticking so as soon as he started firing, he had to then pick up a nearby M-16 rifle and start beating on the .50 caliber to get it to stop firing. Naturally, he couldn't point the .50 caliber very well while he was beating on it, so he was very frustrated with the situation.

I found the scene very amusing. We were picked up by the

wingman within about thirty minutes, and the aircraft was recovered later. We had been shot down with one AK-47 round that penetrated the transmission oil cooler. I have that round fastened to a plaque hanging on the wall in my home.

Below is an email from Jim King, followed by comments from me that reflect some personal feelings.

> Subj: April 14, 1969
> April 15, 2001
> I just had a great conversation with Bob Steinberg about his recent contact with you. I am the same Corporal King who stayed on the .50 cal the day you were shot down near An Hoa. I was trying to provide covering fire to both sides by alternating guns. I realized that the gun I would leave was cooking off rounds after I quit firing and went to the other gun. When the recovery aircraft came in, I was trying to get the butt plates off, as per standard procedure, besides being scared out of my mind!
>
> At a recent reunion, we were trying to identify the pilot of the helicopter I was shot down in on that April morning. So glad to share your memories. Who were the other crew members, if you know? I survived RVN and became an electrical engineer with a mining company. Last year was my first contact with any of the old squadron, it was great. I got to see Colonel Brady, Rich Bianchino, and Ernie Gomez. Rich and Ernie were in a crash we helped recover later the same day you and I were shot down.
>
> It was so good to hear from Steinberg again, he was my first Medical Evacuation Crew Chief. All war stories aside, I hope my message finds you well. Please drop me a note sometime, I'd love to hear more from you.
>
> Regards:
> Jim King—former Sgt USMC, present gray-haired old engineer

Maj. Owen C. "O. C." Baker Remembers
Subject: Re: April 14, 1969
April 15, 2001

Hello Jim King, it is difficult to believe that it was thirty-two long years ago yesterday that we were together in a dry rice paddy in Vietnam trying to stay alive until our wingman could come down and pick us up. Some aspects of that event are as clear as if they happened yesterday. I was delighted to get your email today and to learn of your great success in life after the Marine Corps. I hope your family and friends appreciate the great courage and dedication you showed as a helicopter crewman in the Purple Foxes.

I'll try to answer some of your questions and give you a recap of the incident from my perspective. We had picked up some Marines from out in "Indian country" and got hit by a single AK-47 round as we were climbing out of the pick-up zone.

I'm not sure of the first indication we had of trouble, but the first thing I recall was the Crew Chief (I don't remember his name) complaining over the intercom of all the smoke back in the cabin and the uneasiness of our Marine passengers. I turned my head around to see what the problem was, and I saw seven or eight of the largest pairs of eyeballs I have ever seen in my life as those Marines were peering forward through the smoke toward the cockpit and wondering how soon we were going to explode or crash.

I began an immediate descent toward an island in the An Hoa River we called Football Island because of its shape. It had a reputation for being in bad "Indian country," but I was fairly keen on getting the helicopter on the ground before something bad happened. As we were descending at about seven or eight hundred feet, the bad guys began firing at us. This was a mistake on their part because if they had just waited until we had landed and shut down, they could have had easy pickings. Instead, I added some power, hoping the helicopter would last until we landed elsewhere, and extended our approach to land on the far bank of the river.

We used to fly with "bullet bouncers," heavy, rigid, curved armored shields that covered the front of our upper torso and rested on our thighs. There was a pocket in the fabric on the front of the shield that held a survival radio, and the shoulder harness held the shield against our body. To get out of the seat, one had to first remove the survival radio from the front pocket (it was held in by a large rubber band), unhook the harness, lay aside both the shoulder harness and lap belt, pick up the heavy bullet bouncer, set it out of the way, unplug the microphone and earphone cords, and then turn and crawl out of the seat and back through the opening beside the control closet. It normally took a while to accomplish this exit, and it could not be done gracefully.

Our copilot on that flight was Sam Ware, I don't think he had been in Vietnam very long, and I guess being shot down was sort of a shock to him. I had the helicopter shut down and had gone through the exit process so fast that by the time Sam turned his head toward me to ask me over the intercom what he should do, I was already unplugged and crawling out. (I expect that was a world record for exiting a CH-46, I know it was for me.) I can still remember how amused I was by the look of surprise on Sam's face when he saw he was about to become the sole owner of that helicopter. It didn't take him long to scramble out after me.

Our wingman that day was Lieutenant "Beach" Baldwin. I guess he had to go offload his troops before he could pick us up, but it didn't take that long. Besides we had you on the .50 caliber machine guns and those few Marines with us, so I think we were in pretty good hands during the wait.

Beach has been one of my favorite people ever since that pickup, and it wasn't until years later that people told me a story about him. As squadron operations officer, I had the reputation of being a demanding (read unreasonable) taskmaster among the junior pilots. I never realized the extent of this reputation until I heard this story. Beach was my wingman on another mission (I'm not sure whether it was before or after April 14, 1969) and he was suffering from a case of diarrhea, but he was so afraid of asking me to interrupt our mission long enough to go someplace that he eventually went in his flight suit and then flew for some time after rather than let me know he needed some relief.

You mentioned the crash of Rich Bianchino that same day. That would be the one that killed Lieutenant Mike Nickerson. Mike had been with the squadron since up at Phu Bai and was considered one of the old hands, very experienced and very capable. He was the one who dressed up as Santa Claus in December of 1968 when we flew beer out to the Marines.

I later met Mike's widow and learned he had a very young daughter at the time of his death—so tragic to lose so many good Marines.

Enough rambling from an old man. Let me say again how happy I am to know another one of our Purple Foxes has survived our time together in Nam and is living a long happy life after the Marine Corps.

Semper Fi, OC

Corporal King was one of the very fine young, enlisted Marines that volunteered to be a helicopter crewman in Vietnam. To this day, I continue to ponder the motivation and be amazed by this purely voluntary act. It was dangerous, we suffered many aircraft losses and Marine fatalities (including some that same day in the crash mentioned in Corporal King's email). The enlisted men were under absolutely no pressure or requirement to fly, and the additional flight pay was trivial.

I attribute their actions to the fact that they realized such flying crewmen were essential to the success of the helicopter's mission (unlike jet aircraft where the enlisted stayed back on the ground) and their desire to be with fellow Marines who were actually confronting the enemy.

Their faith in the ability of the (sometimes very young and inexperienced) Marine helicopter pilots up front and their steadfastness in the face of long mission hours and extreme danger during the missions have earned my undying respect for every enlisted Marine helicopter crewman.

INSUBORDINATION
Allan Hemphill

I WAS GETTING PREPARED to go to Prospective Submarine Commanding Officers School in Hawaii, so my Commanding Officer gave me the assignment of making a practice submarine attack on the USS *Enterprise* (CVN-80) during an exercise.

While submerged, I penetrated the screen with our submarine between two American destroyers. There were Canadian destroyers on the screen, but they needed to be avoided because they were *good*. I made the approach on the carrier and a good firing position, and at one thousand yards, we fired three Mark 14 exercise torpedoes directly under the midships of the carrier.

As usual, we fired off a flare to signal our position and surfaced, right into the teeth of a state three sea! It was rough. Watching the bow dip under green water, the Commanding Officer decided he was not going to take the proffered helicopter over to the debriefing. He turned to me and said, "You sunk them, you go tell them." So, with a three-day growth of beard, with the scent of diesel oil and green water on my dirty uniform, I appeared in the wardroom of the *Enterprise*. I was greeted by an audience in well-pressed blue uniforms, complete with ties, and an admiral who was not happy with watching three torpedoes run under his hull.

I explained how we had penetrated the screen to sneak into a firing position, not difficult off San Diego where the inversion layer (a sharp temperature change) is great protection by creating a barrier to prevent the ships from seeing the submarine (the sonar beams bounce off the thermal layer).

"Nice work, Lieutenant, but we were only doing nineteen knots. Would it have made a difference if we had been at forty knots?"

"Not really, sir," I said, matter-of-factly. "At one thousand yards, the *Enterprise* looks like the coast of California." I then swung an air periscope 180 degrees, and said, "Your bow is here, and your stern is way over here."

My Commanding Officer got a message when I returned to the boat—something about "insubordination."

JANITOR TO ADMIRAL
Bruce DeMars

DURING HIGH SCHOOL on the south side of Chicago, I worked as an assistant janitor at the Second Federal Savings & Loan. It was run by the Sierosinski family, and the Vice President, E. John Sierosinski, knew I wanted to attend the Naval Academy. He called me in during my senior year in high school and told me I had an appointment at the Academy but had to use this address—46 Northlake Road, Riverside, IL—a very affluent suburb. I thanked him and asked if there was anything else I should know. He told me that the Riverside congressman owed the editor of the *Southwest News* a favor, the editor owed him a favor, and I was the favor. However, the congressman was on record that the appointment would be competitive, so I had to go downtown to the post office building and take the civil service exam. There were others there from his district, and they did think it was competitive. I, of course, said nothing.

I received the appointment but failed my physical at Great Lakes because of high blood pressure. I traveled to Annapolis by train and bus for a re-exam and arrived late at night at the Visiting Team Barracks with the others. The next morning, a corpsman took my blood pressure, and it was still too high. He said I looked tired, which I was, and told me to lie down on a bed in one of the wards. I fell sound asleep and was awakened by the same corpsman, again taking my blood pressure. He said, "Red, it's okay, you're in." So, I started my long career on shaky legal grounds but with an enduring respect for the fine judgment of enlisted men!

A Pentagon Caper

In 1987, I was a three-star admiral on the Chief of Naval Operations staff when I was appointed to be president of the selection board for Captains. I called on the CNO, Carl Trost, and he told me to do a good job. I called on the Secretary of the Navy, John Lehman, and he gave me a list of people he expected to be selected! I found this strange but didn't say anything. We had a good board and worked hard to follow the precept issued by Secretary Lehman. We selected some of the Lehman names, on merit, but not all. I found that some others on the board had the same names.

When I made my out-brief list for Secretary Lehman, he was quite upset I hadn't done exactly what he wanted. I explained it wasn't legal, and he had communicated to us a precept we had rigorously followed. I reminded him that each flag officer on the board had sworn an oath and signed the selection report. This started about two weeks of regular meetings with him and selected members of his staff to get me to do something different. I kept the CNO informed as we went back and forth and knew I had his full support. Lehman even offered to give the board five additional selection numbers if we would reconvene and select those we wanted. I refused. He said he was going to remove me as president of the selection board. Hearing that, I wrote a letter of resignation that somehow got out to the press. I began to consider a civilian career.

The pace now really picked up. Lehman portrayed the matter

as civilian control of the military and had several others write articles and op-ed pieces supporting him. I received many letters and phone calls supporting me. The Secretary of Defense Inspector General was called in and all were interviewed. The report upheld my position and about a month later, Lehman resigned.

Epilogue—About five months later I was nominated to be the Director of Naval Reactors. Lehman started a rumor that the influential Senator Tower didn't like DeMars. A senate staffer friend of mine called Tower, who was in London, to ask if he supported DeMars. Tower said, "Who is DeMars?" and the matter was over. I was a Four-Star Admiral.

SUBMARINE ICE EXERCISE
Tommy Sawyer

FIRST, I HAD A TOUR OF DUTY in the Torpedo MK-48 Project Office during the Torpedo contractor selection process. Then I returned five years later as Test Director just in time for the test and evaluation phase of the first major improvements of the torpedo. The performance of the torpedo against both surface and submarine targets under normal environmental conditions exceeded expectations. However, there remained the important unanswered question of how the torpedo would perform in the under-ice environment. If the Cold War turned hot, this is the environment in which it probably would be employed.

Planning for such tests was put in motion, and I was designated the on-scene Test Director. We needed a site where the ice would support test personnel and equipment but also permit exercise torpedo recovery. However, it had to represent a real playing field in which to test both acoustic and warhead sensors. We had to know if the torpedo could distinguish between real targets and ice. With the assistance of our torpedo, undersea, and arctic labs, a site was selected. It would be closer to the North Pole than to the Arctic Circle. Just the thought of working in that environment was daunting.

The actual torpedo firings to answer our question would be relatively easy. One of our newest fast-attack nuclear submarines trained in under-ice operations would be able to complete this task in a few days. It was the planning and logistics before and after the actual torpedo firings that were the hard part.

The first order of business was to construct an ice camp to support the operation, and prefabricated building materials were airlifted to the sight. The heart of the camp was the Operations Control Center, which included a communication center and tracking equipment for both submarines and torpedoes. The camp provided berthing and a mess hall for thirty people for about forty days. As one would imagine there were a lot of scientific projects along for the ride.

An under-ice instrumented tracking range was constructed to know the position of the submarine, torpedo, and target simulator. This required positioning of range hydrophones over a large area below the ice cap. When the torpedo had completed its run, it would float to the underside of the five-foot-thick ice. Its approximate position could be determined from the tracking range, and a specially designed ice-cutter was used to cut a four-foot diameter hole in the ice. A SEAL team diver would then pull the torpedo to the ice hole where it would be winched to the surface or lifted out by a helicopter. The helicopter would transport the torpedoes to an intermediate staging area before being air-transported home for a complete post-run analysis.

The planning and execution of the operation included personnel and equipment from three countries, fourteen US Navy commands, and six US Air Force commands. All torpedoes were recovered and provided the required information for employment of our torpedoes in the under-ice environment for present and future generation weapons.

It was an exciting experience. Fortunately, we didn't have to use our "polar bear encounter" training. Just existing where the average temperature was minus twenty-five degrees below zero was challenging enough.

PROJECT SUBICEX Pictures with Tommy Sawyer and a Mark-45 Torpedo.

THE GREAT AND GRAND TEAR GAS MISSION
Charlie Hall

THIS ADVENTURE STARTED OUT with some tales my guys gathered from a guy who knows a guy! They came to me with an idea, and we hatched a plan and then an Operations Order. Where it started, one of my guys came to me and told me he heard, from a reliable source, that tear gas (a) comes in droppable canister bombs and (b) if they were bombed in a known Viet Cong hangout, they would endure the discomfort and stay underground or in their bunkers until the gas cleared, but if they were above ground, they would not be able to get into their caves or bunkers due to the tear gas debilitating their senses, and so they would be targets for small arms fire we carried on the helicopter.

So, that led us to go talk to our US contacts in our Operations Areas and zero in on known VC living areas. We chose one such as a target: a wooded and brushy area reported to be concealing several bunkers and hootchies occupied by the local VC. We then set about a glorious assault on the bad guys when and how they might least expect it. One of my pilots went after the tear gas canister bomb and the instructions on how to deliver it manually from a Huey helicopter. Another worked on the tactics for the assault flight. And still, another worked on the weather conditions for the next few days.

Then the plan evolved as follows: First, we needed to get clearance from the proper Republic of Vietnam authority, which we did through our US Army Commander in the chosen area. That was not hard at all, and so we moved ahead. My guy came back with the ordnance and the instructions. He briefed us on how to deploy manually and that part seemed simple.

So now we had to check the weather for wind conditions. We wanted winds light and *steady* since we wanted the gas to stay put on and near the ground where we deployed it. And that was the forecast for the following day in the area! Tactically, the lead Huey would fly a crosswind leg at about two thousand feet just upwind of the target area and deploy the weapon at about the center point of the target. The trailing Huey would be prepared to provide cover fire.

Then both Huey's would set up in an oval pattern, like a holding pattern. Once the bomb detonated and the gas dispersed, the gunners would put in fire as the target area bore in their sights. That way we expected to (a) catch the VC above ground and keep them there and (b) deliver considerable firepower on them while they were unprotected.

Now it was getting close to Go-No-Go-Day. I decided to lead the flight and chose one of my most experienced Helo Commanders to man the trail Huey. The morning came and still everything looked fine for this mission that was designed to bring death and destruction on hordes of the VC enemy who deserved anything we could inflict upon them. Hell, hath no fury like a couple of fully armed Hueys with tear gas canisters and enthusiastic gunners. So, we launched, called the appropriate US Liaison Officer, got our clearance meaning that no friendlies were in the area to interrupt our plans, and proceeded to our mission area. I identified the target area, and we proceeded to set up the drop. My gunner would unsafe the weapon and arm the radar fuse to detonate at about two hundred feet. That would allow the major weapon to deploy the submunitions as the canister fell the rest of the way to the ground.

So here we came, the Armored Huey Detachment! The day was CAVU (Clear View Up) to the moon, the winds were light and steady, and we approached the target from the upwind side. I flew the crosswind leg, and my Head Gunner deployed the tear gas canister on my command. We watched it fall and the trail ship called detonation as they could see the white cloud indications. The smaller munitions scattered out from the canister just as advertised, and all was going well! We were just a moment away from dealing death and destruction to our enemies! We set up in the oval pattern just upwind of the target and got ready to open fire.

But as a colleague once told me, "Nothing ruins an operation like the order to execute!" Just as we began our first firing run, the dastardly wind shifted almost 180 degrees and picked up a bit. Suddenly, the gas was blowing in our direction and, since it is colorless, we had no warning and both ships got a huge dose of tear gas.

Now, I have never been in a demonstration-turned-riot, where tear gas was employed to control the crowd, but I became aware of the effects of a dose of tear gas in a hurry. So did all the other crew members of both birds! I had trouble breathing, my eyes were stinging, and my orientation skills were debilitated.

All that in a New York second (the time between the light turning green and the cab driver behind you honking his horn!), and I was immediately more concerned with maintaining control of my bird than doing anything about any assault on the enemy!

But, after a few very scary moments, things began to clear up, and I was okay again, just some nasty, stinging eyes that kept me blinking quite a bit. I was in control of my bird, my trail ship commander was in control of his bird, and we were headed home—a bit disconcerted about our failure to execute our planned mission and wondering how the wind shifted so fast. Damned weather-guessers! Wind and weather, as all pilots know, can be very devious and tricky, especially when you least expect it. We decided we would not try that plan again, but we all thought as often is the case after events like that and also after a liberty in a strange port, "At the time, it seemed like a pretty good idea!"

THE LONDON GARDEN PARTY
Jack Homnick

ON OUR SUMMER CRUISE after our first year at Annapolis in 1954, Bob Mazik and I were assigned to the famous battleship USS *Missouri* (BB-63) (where the Japanese signed the papers to end WWII). When the ship docked in Cherbourg, France, American Express sold us a four-day excursion trip to London. Bob (a.k.a., Hunky) and I decided to go there while most of our other classmates hopped on a train to Paris. We flew a chartered Air France from an abandoned Nazi airstrip (grass growing up in the tarmac cracks) on the Normandy coast to Heathrow. A classmate and Missouri shipmate had a relative who worked at the American Embassy, and he got us invitations to the American Ambassador's Fourth of July Garden Party. Here we were, measly Midshipmen, going to *the* American party of the year in London!

We splurged and took a cab to the Ambassador's residence (not near the embassy) and entered the front hall where an attendant took our caps and placed them on a shelf along with many Admirals' and Generals' hats. We dutifully placed our calling cards on the hall table and were ushered down the long hall to the rear porch area where the Ambassador and Mrs. Aldridge were receiving guests. As we approached, a voice announced, "Midshipman Om-nik, Midshipman Mah-zik." We were very graciously received and made to feel welcome and shown the way to the party in the luxurious gardens behind the residence.

And what a party it was! I think we were the only Midshipmen there and certainly, the only young bucks as we were introduced to many lovely ladies and daughters of attendees. The spirits flowed very freely, and it was a beautiful summer evening, highlighted by the playing of the "American National Anthem" and "God Bless America," sung by Kate Smith (?) or someone who looked like her. We met bearded Admiral Pirie who was the NATO chief at the time and many, many other admirals who took us in tow and delighted in telling everyone we were the "new Navy." We also met actor Douglas Fairbanks Jr. and Sonia Henie (the ice champion/actress). We had a glorious time and got completely smashed—as were all the admirals as I recall. At some point, the spigot got turned off and everyone left, sadly, end of the party.

Bob and I got a cab back to our favorite Piccadilly pub for a nightcap. The barmaid called us "My Yanks" and had some brews on ice for us as we had been there several nights before. She always laughed at our wanting iced beer. I don't think we ever paid for a drink . . . God love the Brits! I've never forgotten that lovely event in July 1954.

THE SAFE
John Stacey

IN 1964, the Navy was in the middle of a submarine-building boom and our goal of forty-one ballistic missile submarines were being turned out at several shipyards, both civilian and military. I was at Newport News Shipbuilding and Dry-Dock Co., building/outfitting USS *John C. Calhoun* (SSBN-630). With many subs being built at Newport News, Virginia, we were packed into offices at the available buildings about a half mile from the waterfront.

The captains of the boats recognized that if we were closer to the subs, we would be more efficient and not waste so much time walking to the boat from the office area to sign off the completed work by the yard personnel as required. When the yard completed a job, they looked for a boat officer to check it out and sign off a completion statement. Soon several two-level barges were brought alongside the submarines at the outfitting piers, and we moved out of our offices and into offices on the barges.

These barges were last used during WWII and had been sitting in a reserve fleet. As Communications Officer, I had a large wire cage as an office, and inside, was a large double-door Mosler safe, perhaps eight feet across and five feet high. This had been a Communications Officer's safe during the war, and we looked around unable to find any combination for the locked safe.

Finally, I called the locksmith at the shipyard who came down to survey the situation. He pulled out a stethoscope and started doing his thing. Within about five minutes, he determined the combination and wanted me to open the safe door. To my surprise, I found forty to fifty code books dated the summer of 1945, classified as top-secret cryptographic material. WOW! What a find! Now what?!

The locksmith showed me how to change the combination, so I set the new one and locked the safe. Then I consulted the "bible" on crypto material to figure out what was applicable here. This situation was not covered. I advised my captain and told him the code books were safely secured, and I was researching the problem.

The next day I called the Naval Security Station, the mother of all crypto affairs, on Nebraska Avenue in Washington DC, and talked to several individuals with no results. No one had an answer on what to do. Everyone promised to get back to me soon. Days went by with no return phone calls. The captain directed me to send a message to see what would happen.

More days went by, and no one answered the message. Finally, the captain directed the material to be inventoried and transferred to Atlantic Submarine Headquarters. I dutifully complied and took the package over to my old buddy, Lieutenant Dick Dolliver, the Assistant Staff Intelligence Officer. He thanked me and accepted custody of the material, and I was on my way back to the *Calhoun*. I talked to Dick later and, since he never received any guidance from above, he finally burned the material and sent a destruction report to the Naval Security Station and never heard anything more.

Apparently, at the end of the war, the Communications Officer received his orders to return to civilian life and did just that. It appears that he closed the safe, locked it up, and walked off. The contents of the safe had been there for almost twenty years, securely stowed in a thousand-pound Mosler safe on a Navy Barge in some reserve fleet location. Top secret code books, safe forever!

THE STOLEN WALLET DRILL
Bob Strange

AS A BACHELOR assigned to a Pearl Harbor-based destroyer, I decided one Sunday to drive to NAS Barbers Point for a little surfing. At the time, they had an officers' beach house for your clothes while using the beach. Finding no place to secure my valuables and since it was after all an officers' beach house, I left my wallet with my clothes in the dressing room. After a couple of hours of surfing, I returned to the dressing room to get dressed and discovered my wallet was missing.

I returned to my ship and had no problem getting back on base with my base sticker on the car without showing my ID. During dinner in the wardroom, I received a phone call from someone at Barbers Point who had found my wallet on the side of the road with everything intact except for the cash and that it would be left at a certain location where I could claim it. I returned to NAS Barbers Point and unbeknownst to me, security had been increased at the base as there was an ongoing security exercise in which a planted person would attempt to penetrate their base security.

I was stopped at the gate to show my identification. I advised the sentry of the circumstances of my stolen wallet. Of course, I immediately became the suspected person and was directed to park in a designated area to await the arrival of the Security Officer. After a prolonged period, he arrived, somewhat irritated at having his Sunday plans interrupted. I explained the situation and was then directed to follow his car onto the base. After several turns, we parked at the NAS brig.

Again I explained I had nothing to do with the security drill and that I was only there to reclaim my wallet. After several phone calls my story was confirmed. My wallet was found, and I was released. The lesson learned: officers' beach houses are no more secure than any other unlocked ones.UGH!

WOO POOS TRIED TO KILL MY WIFE
Jim Paulk

MIDSHIPMEN AT ANNAPOLIS called cadets at West Point "Woo Poos" with smiles on their faces. The Army/Navy football game, a classic, has generated a friendly rivalry over decades between the two schools. During our senior year, we went to West Point to spend a week with cadets, while cadets spent a week at Annapolis. There was also a reciprocal exchange of officers where an Army officer spent a tour of duty at Annapolis and a Naval officer went to West Point, each serving as a Company officer for a company of cadets or Midshipmen. This interaction between the two service academies not only created a natural rivalry but mutual respect for the experiences at a military academy.

One of my best friends from Brunswick, Georgia, Carl Croft, had a lifelong dream to attend West Point, and I had a similar dream to attend the Naval Academy at Annapolis, but neither of us received an appointment after graduating from Glynn Academy (our high school). So, we both went to North Georgia

College in Dahlonega, Georgia, located in the mountains. We chose this school because it was an accredited military school with excellent academics and we could hopefully be admitted to a military academy the next year.

After a year at North Georgia, I did not receive my appointment to Annapolis, but fortunately, Carl was successful, and he entered West Point. I went back to North Georgia for a second year and finally received my appointment—my dream had come true!

We made bets on the annual Army/Navy game with friendly taunting over the telephone, but I'm getting ahead of myself.

On the day I graduated from Annapolis, my wedding party celebrated our graduation, and the forthcoming wedding was planned for the next day. The party was at the Media Inn, south of Philadelphia. Needless to say, all of us had too much to drink and had trouble finding the house arranged for our stay, but we all eventually found a bed, closet, couch, or floor for a few hours of sleep. With Carl representing the Army, we had all four services in our wedding party.

Since we came from a small town and had known each other all our lives, Carl and I stayed in touch over the years and visited as opportunity permitted. His story is an interesting one and the genesis for this story.

After graduating from West Point where he played on the Army football team, he was stationed at a missile battery base near Philadelphia, but after a year, he decided that career path was not right for him. Somehow arranging an appointment with the Army Chief of Staff, General Maxwell Taylor, he negotiated a leave of absence from the Army to attend medical school at Duke University. After med school, he went back into the Army and was sent to the Presidio, an Army base near San Francisco, for his internship. Upon completion, the Army posted him to West Point as a physician and football team doctor.

In the meantime, I completed my first sea duty on a troop transport, USS *Montrose* (APA-212), applied for and graduated from submarine school. My first submarine was the USS *Redfin* (SSR-272) where I was assigned as the Electronics Officer. With this assignment, I was sent back to New London for a six-week course in submarine Officer's electronics training. On one drive to New London from Norfolk, I dropped my wife, Pat, at a New York City airport as planned for pickup by Carl, and his wife, Ann, to have an extended visit with them in their quarters at West Point. Days later, the plan was for me to retrieve her at the same location. This was a good plan, what could go wrong?

With no cell phones in those days, there was no communication between Pat and me while she was at West Point. Why should there be? She was in good hands with friends. When I returned to the airport to pick up Pat, the three of them were standing at the curb, but something didn't look right. Pat had an arm in a sling and assorted bandages. What had happened to my beautiful wife whom I had left in the care of good friends? Had the Woo Poos there tried to kill my Navy wife? I was in shock!

Of course, Carl and Ann were concerned too about the accident that occurred at their quarters. The parking area for cars was at a higher level than the entrance to the living quarters and there was a flight of stone steps there. After a shopping trip, both wives returned home, and Pat's high-heeled shoe became stuck in a crack at the top of the stairs, and she tumbled down stone steps. Fortunately, Pat's athletic training came back, and she relaxed during her fall, or her injuries could have been worse. There were no broken bones, only bruises, cuts, and embarrassment. When I first saw her, I was scared to death and asked her if the Army people had tried to kill her, but over the years we laughed about this incident which thoroughly frightened Carl and Ann.

After Carl completed his obligated service with the Army at West Point, he and Ann moved to Winter Park, Florida, where he was an orthopedic surgeon for thirty years and the team doctor for the Orlando Magic (Shaq O'Neal's team) and college and high school teams in the area.

TAKING COMMAND
Jim Poole

IT WAS SEPTEMBER 1973, and I was busily engaged in the process of assuming command of USS *Norton Sound* (AVM-1). *Norton Sound* had been commissioned near the end of World War II as a seaplane tender and recommissioned after the war to provide a sea-going platform for the testing of naval weapons and weapon systems being developed by the then-Bureau of Naval Ordnance.

As I assumed command, *Norton Sound* was to complete testing of the MK 26 Guided Missile Launching System, after which it was to proceed to the Long Beach, California, Naval shipyard to install the first engineering development model of a new weapons system. Upon completion of this installation, *Norton Sound* was to return to her homeport at Port Hueneme, California, for an extended period of at-sea testing of the new system, including the live firing of standard missiles against a wide variety of system stressing targets on the Navy's Pacific Missile Test Range at Point Mugu, California.

Although *Norton Sound* was administratively attached to an afloat command based in San Diego, California, we operated quite independently because of our mission and reported operationally to the new Program Manager attached to the Navy Ordinance in Washington, DC. At the time, that Program Manager was Captain Wayne E. Meyer. Although I was not to learn of it for several years, Meyer had someone specific in mind to take command of *Norton Sound* and was quite upset when I received those orders to do so. He tried to have the orders modified, but, fortunately for me, the Bureau of Naval Personnel turned a deaf ear to him.

Captain Meyer was, to put it mildly, a unique individual. Well known today as "the father of this new weapons system," but to some, he was the epitome of the renowned Admiral Hyman G. Rickover, father of the nuclear Navy — arrogant, domineering, explosive, and unyielding. To others, he was a brilliant engineer, intense, loyal, dedicated, energetic, and committed. Charlie Clynes, the man I was relieving of command of *Norton Sound*, had little good to say about Captain Meyer however, and advised me to "ignore him; he doesn't write your fitness report." (Oh, yeah?) To my everlasting credit, I ignored that advice and decided I would make up my mind about Wayne Meyer.

The day before my Change of Command ceremony, I was advised that Captain Meyer, in company with his Technical Director, Captain Paul L. Anderson, would arrive on board the following morning at 0800 and wished to speak with the new Commanding Officer. I had never met either of the captains and, after Charlie Clynes's glowing endorsement of Meyer, I was somewhat apprehensive of the meeting, knowing full well it would be one of sizing me up and telling me what was expected of me.

The next morning, promptly at 0800, Captains Meyer and Anderson arrived. After introductions on the quarterdeck, we proceeded to the flag cabin where, as I had guessed, Captain Meyer told me of the importance of the new program to the Navy, how much would be expected of me and my crew, and how our performance would be graded in only one of two ways—"either A-plus or F-minus." I responded to Captain Meyer by saying how pleased I was to have the opportunity to command *Norton Sound* and to embark on a test program of such significance. I told him most of my career had been involved with weapons and weapon systems of some kind, and I was not unfamiliar with testing and test programs. I concluded by assuring him that I and my crew would give the ship, him, and the program 110 percent effort. We concluded our conversation at this point to prepare for the ceremony.

The Change of Command proceeded without incident. After relieving Commander Clynes and reporting my relief, I took the opportunity to make very brief remarks which I addressed to my new crew. I expressed my pleasure in assuming command, told them we had a very challenging and extremely important mission ahead of us and ended by saying we would be graded in our performance in only one of two ways—"either A-plus or F-minus."

Two years later, after I had reported to the new systems program office in Washington for duty, Paul Anderson revealed to me at that point in the change of command ceremony when I made those remarks, Captain Meyer leaned over to him and whispered, "Learns quickly, doesn't he?"

SOFTBALL GAME
Joe Koch

ON OUR WAY INTO the Mediterranean in June 1960, our submarine, *Tusk* (SS-426) made a liberty port call at Oporto, Portugal. We were the only American warship in port at that time, and there had not been an American ship there for several years before us. At that time, Oporto had a rather large ex-pat British colony since most of the local wineries were managed by Brits, and since there was not an ex-pat American group in Oporto, we were adopted by the Brits for our stay in port.

One of the events the Brits arranged for us was a softball game on Sunday afternoon at the Oporto Lawn Tennis and Cricket Club. It was the *Tusk* crew against the Brits. However, they insisted we play by "British rules" which differed from our rules only by a "tea break" after four and a half innings.

When "tea break" came, *Tusk* was ahead comfortably by a score of twelve to four. "Tea break" lasted an hour and a half, and only strong British beer was served instead of tea. Luckily, the Brits were imbibing as much as the "Yanks." *Tusk* managed to squeak out a thirty-six to thirty-four victory after nine innings. That was the wildest second half of a softball game I have ever participated in. However, everyone had a great time no matter who won. I always wondered why we did not have more softball games using "British rules!"

1964 THRESHER SEARCH
John Howland

WHEN I FINISHED MY POSTGRADUATE COURSE in oceanography at the University of Washington in Seattle in 1964, I received orders to the bathyscaphe *Trieste II* stationed in San Diego. The first *Trieste* received a new float—hence the new name, *Trieste II* (the compartment for the pilot and copilot was ball-shaped and suspended under a larger submarine-shaped vessel called the float).

We had been ordered back to the Atlantic to conduct a search for the hull of the US *Thresher* (SSN-593) that had been lost in April of 1963. An original search had been conducted in 1963 and identifiable pieces had been retrieved from the debris field, but the hull of *Thresher* had not been found. We were sent back to find it.

Shortly after reporting aboard, I got the job of accompanying the *Trieste II* (DVS-1) back to Boston on an ancient World War II Liberty ship by way of the Panama Canal. The ship was not air-conditioned, and my stateroom was on the top deck where the sun beat down directly on my overhead (roof)—not a fun trip!

Upon our arrival in Boston, we were assigned a submarine rescue ship as a tow ship and the US Naval Ship *Mizar* (T-AGOR-11), which had been equipped with navigation equipment and deep cameras, would accompany us.

Being the junior guy, it was decided I would make every dive as copilot in the *Thresher* area for continuity purposes, with Brad Mooney, USNA Class of 1953, and Larry Shumaker, USNA Class of 1954, alternating as the chief pilot. The underwater navigation equipment we had was primitive, at best. We made several dives in the area and found only more debris.

However, we made a dive on August 17, 1964, in the designated area where the debris from the boat had been located the year before, and after no results for several hours, we settled on the bottom in about eight thousand feet of water to reorient ourselves. After about thirty minutes, we received a direction from the surface to steer and lifted off the bottom to go where we had been told. Brad Mooney was at the window forward, and I was keeping track of the underwater television cameras that were mounted amidships.

As we lifted off the bottom to start moving, suddenly, the picture on one of the television monitors became clear, and I told Brad to stop the ascent. We were on the hull of the *Thresher*! We had been sitting on the hull for thirty minutes trying to figure out where to go to find it. Had we not stopped to reorient, we might have missed it, as it was covered with a layer of silt and was not obvious when just passing over the top of it.

The Navy's Deep Submergence program was spurred on by the experiences and discoveries we made that summer, and resulted in, among other things, the development of the Navy's Deep Submergence Rescue Vehicles for submarines.

FIRE DRILL
Bob Christenson

I WAS THE EXECUTIVE OFFICER of the submarine *Baya* (SS/AGSS-318), and the crew was honed to a razor's edge! It was time for a real drill. With the Captain's permission, I went up to the Squadron Office, saw the Training Officer, and asked for an oil smoke grenade, commonly used to simulate smoke from a fire. Proudly bearing my secret back to the boat, I locked it in my desk safe.

A week or so later, during a most demanding training cruise, we were demilitarized and used for sonar research down the coast of California on the way to a hectic schedule in Mazatlán, Mexico. It was time for the true test. The boat was at a depth of one hundred feet. The Captain said, "You sure this is going to work, XO?" I assured him it would be a piece of cake!

Into the radio shack, a small closet-sized room just off the main control room, I snuck the bomb. I told the Chief Radioman to pull the pin and drop it in the empty waste basket in a minute or two so I could get into the forward battery compartment from whence

I could jump into the control room and observe the razor-sharp crew perform in their normal confident and cool manner as we surfaced and simulated fighting the fire.

Soon, all hell broke loose! Indeed, we had smoke, gagging, coughing, choking, and other signs of great respiratory distress. We got to the surface fine, but the ostensibly benign oil smoke was more than it should have been. After getting great numbers of the crew topside into the fresh air and thoroughly ventilating the boat with the diesel engines running to use up the toxic air, things finally calmed down—more or less.

We put on speed for Mazatlán and sent a message, designated "Immediate," to the squadron, reporting the designation, serial number, etc. of the offending training device and requested the services of the squadron Medical Officer in Mazatlán to check the health of those who seemed to have had the greatest exposure.

Within a day we had pulled in, and there was our Medical Officer waiting on the pier, surfboard in tow! After extensive examinations of all who had been subjected to the fumes—with one exception, who needed a chest X-ray—this smart-a-- medical officer prescribed lots of rest, sunshine, and rum to ward off any lingering whatever. We reluctantly obeyed his advice, giving up the visits to local art galleries, churches, historic plazas, gardens, etc., which is why we had chosen this port in the first place. Not for real folks!

All is well that ends well! Well, not quite. A message from the squadron advised us that the smoke grenade was not an oil smoke but a marker smoke, used to designate landing zones for helicopters and other goodness-knows-what uses. Turns out, it was used by the squadron training officer to mark his descent as he *jumped out* of the aircraft to parachute to the ground! And he was a submariner!

Silly bastard . . . he had given it to me by mistake. The message also advised us that upon return to San Diego, there would be a Squadron Commander's Inquiry (like a court hearing)—THE LONG GREEN TABLE. *So much for my naval career*, I thought.

However, rested and relaxed after our arduous duty, we returned to home port and after some posturing, the Squadron Commander felt my only negligence had been to not turn over the offending missile to the Torpedomen, for storage in the pyrotechnics locker, who might have noticed what we had. After a summary slap on my wrist, he said, "There, but for the grace of God, go any number of us."

I never had the privilege of commanding a submarine for reasons entirely different from this incident, but hail those who did. My career went in a different and very satisfying direction, but I never forgot that everyone, officer and enlisted, in most cases probably deserves a second chance. Thanks, Squadron Commander!

A RESCUE AT SEA
Bob Wellborn

THIS IS A STORY about providence over the lives of those who go to sea. The story is true, but I tell it so each of you can draw your conclusions about providence. One first-person account of the story was carried in the Pacific *Stars and Stripes* issue of April 10, 1969. The title is "Lucky Sailor Glad to Be Aboard—Again." That story is about Machinist Mate Second Class Jim Davis, who washed overboard from the USS *Porterfield* (DD-682) while attempting to rescue a fellow sailor during a storm off the coast of Japan a few days earlier. Here is the "back story."

I was the Executive Officer in Porterfield, one of the last of the five-gun Fletcher class destroyers, valued for our ability to fire simultaneous multiple gunfire support missions during the Vietnam War. We were returning home to San Diego from Vietnam along with the destroyer USS *John Paul Jones* (DD-932). First, though, we had to escort the aircraft carrier USS *Midway* (CV-41), to her homeport in Yokosuka, Japan.

Several hours before sunrise, on the morning we would arrive in Yokosuka, a vicious storm suddenly came up while we were heading north in the outer approaches to Tokyo Bay. We were about sixty miles off the east coast of Japan. The storm drove steep, choppy waves across our path with a strong wind coming off the Japanese island of Honshu to our west. Estimated wave heights became twenty feet, with fifty-knot winds off our port beam. This was particularly uncomfortable for our salty old destroyer. She did not have protected passageways inside the ship for people to move between watch stations, berthing spaces, and the mess deck. (If you would like to get a mental picture of this, watch the Tom Hanks movie, entitled *Greyhound*.) Our heavy weather procedure was to allow only essential personnel movement. We required everyone outside to be above the main deck, on the first level (level above the main deck), or the torpedo deck as it was known in WWII when they left their living compartment for meals and watch standing. When we expected bad weather, the bosun mates rigged an extra set of inboard lifelines on the main deck. These still would not help if a man was hit full force by a storm-driven wave breaking over the side of the ship and onto the main deck. Most men wouldn't have the strength to hold on against both the force of the water and its breathtaking cold shock.

This storm was unexpected, and we didn't want crew members floundering in the dark. Therefore, we decided to wait until daylight to put the ship on a safer course and then rig the heavy weather lifelines. The newer and larger *John Paul Jones* could handle the choppy seas better than *Porterfield*, and the *Midway* was too big to realize much of anything was happening. The carrier, like a battleship, pitches and rolls in large swells like those off Cape Hatteras, but not in the twenty-foot choppy waves we were enduring. Even so, every carrier is solicitous of sea-keeping conditions on smaller escorting destroyers. Slowing and changing the entire formation to an easier riding course for the smaller ships is the typical storm response.

All of us on this particular morning were willing to trade a little discomfort to get into port sooner for a long-awaited liberty call. The sunrise discussion with the Captain, myself, and the Officer of the Deck was whether or not to run downwind and delay arrival while rigging heavy weather lifelines. Or we could keep everyone off the main deck until we moved closer inshore and out of the storm. We decided to delay breakfast while warning everyone of heavy seas breaking on the main deck. We planned on reaching calmer water by about ten a.m. The plan did not work.

Events on the ship took their own course, but descriptive information is required first to picture the situation in *Porterfield*. The propulsion and electricity generating system on a Navy ship is called the engineering plant, or just "the plant." Those who operate and maintain it are called engineers. *Porterfield* has four oil-fired boilers divided into two firerooms. The boilers power two steam turbine engines, each in a separate engine room. The separation of the plant into four watertight spaces provides both redundancy and isolation in case of fire or flooding in one space. The only access to each of the four engineering spaces is by a steel ladder coming down from a watertight hatch on each side of the main deck. If heavy seas are breaking more on one side of the ship, the engineers use the hatch and ladder on the opposite side. This prevents seawater from washing down through the hatch and also protects the engineers from being carried away while opening the hatch. In truly bad storms, the engineers stay in place for the duration.

The morning watch relieved the midwatch by four a.m., about the time the ship began to roll from the storm. The Oil King, the man responsible for pumping and ballasting all the fuel tanks, came on watch shortly after four a.m. He began his last round of soundings (gauging levels in the tanks) before entering port. Soundings on the four service and eleven storage tanks, located throughout the ship, enabled him to prepare an accurate estimate of how much oil we would need to fill every tank. As soon as a

Navy ship ties up in port, the oil barge comes alongside, and the engineers spend the next four hours refueling the ship. Then she is ready for sea in an emergency, as well as fueled for the next leg of the voyage, and only then can the engineer's liberty section leave the ship.

Despite the dangers on deck, a storm at sea is fascinating. Constantly rolling and pitching over wave after wave, our old destroyer won every battle. Her bow dug in and then lifted green water from each defeated breaker, sending it careening harmlessly down the main deck and back over the side. Safe on the bridge, two levels above, we thought salty bulkheads, some extra rust, and tarnished brass from flying spray would be our only price to pay for this battle with the sea. That thought vanished in the next instant.

By the six-a.m. sunrise, the Oil King completed his soundings (determining oil in the tank) and climbed the ladder, away from the storm's side of the ship, from the after-engine room to the main deck and then on up to the Log Room on the first level. There, despite the rolling and pitching caused by the storm, he planned to compute barrels of fuel from the feet and inches of his soundings.

The Oil King reached the top of the ladder, clipboard in one hand, just as a rogue wave's defeated remnant slammed into the bulkhead below him. Reacting to keep his balance, and save his clipboard with two hour's hard work, the Oil King lost both. He fell to the main deck and was carried like a surfboard along the aft deckhouse, the aft gun mount, onto the fantail, and into the depth charge racks. Two early risers, one of them the ship's Hospital Corpsman, the other a young engineer, Machinist Mate Third Class Jim Davis, were standing in the log room door as the Oil King fell and lay stunned while being swept aft to the depth charge racks. Their reaction was immediate; save the Oil King. They leaped down the ladder and ran aft, reaching the Oil King just as a second wave surged down the fantail deck. This one swept the Hospital Corpsman into the same depth charge rack, jamming his leg in the angle iron foundation and pinning the Oil King underneath by his head. The other selfless rescuer, Jim Davis, was swept overboard into the Sea of Japan.

The after lookout, connected to the bridge by the indestructible Navy "sound-powered telephone," immediately reported the Oil King's fall and then the unsuccessful attempt to rescue him. Within the minute, the dreaded "man overboard," announcement came over the ship's general announcing system. Response to "man overboard" is practiced over and over on a Navy ship until every watch stander is trained. The Officer of the Deck and the Combat Information Center began the rough weather recovery maneuver and tracked the relative position back to where Davis went overboard. The engineers began setting up boilers, superheaters, pumps, generators, and main turbine throttles for maneuvering and backing down to stop the ship when we approached him. Life rings and markers went overboard to keep him afloat and in sight. The rescue team donned wet suits and rigged tending lines to aid in bringing him back aboard. The lifeboat crew went to their station, even though rough seas made its use unlikely. The carrier ordered *John Paul Jones* to aid us in the search. *Midway* was too big to maneuver effectively in this kind of situation. That is why we were escorting her. And her helicopters were down for maintenance since no flight operations were planned.

Under normal conditions, our trained reactions could have quickly initiated Davis's rescue. The storm conditions disrupted the timing we achieved by teamwork in man overboard drills. *Porterfield* couldn't make the man overboard recovery maneuver, called a "Williamson turn." First, we came to a safe course while the recovery team rescued the Oil King and the ship's only Hospital Corpsman from the depth charge racks. Heavy rolling and waves breaking aboard hindered everyone trying to man their rescue stations and get aft. Holding the safe course for several minutes allowed them to get in place and complete the rescue on the fantail. The terrible loss of one man overboard in a storm could not be compounded with choices causing others to be lost as well. The odds of finding Davis already were low, so there was no point in making the situation even more tragic by losing more lives.

As we finally came fully around to the southerly course to get back to Davis's estimated position, the only thing we could see was blinding spray whipped off the tops of the whitecaps. We now had the storm's effects attacking us on our starboard side as we "rolled our guts out" and plowed ahead into twenty-foot waves. The bridge window wipers were no more effective than those on a car during a heavy downpour. The men on the open bridge wings spent much of their time ducking as spray came slashing over the side. Our ship was too small to see over the storm. Even the signalmen on the third level above the bridge were hampered by the sheets of spray. "Sea return," multiple echoes from the waves, filled the radar scopes masking the marker beacon. Sheets of spray drenched the lookouts, making their binoculars useless except for the few seconds after they cleaned them. The ship's hard-to-predict drift from the high wind velocity added inaccuracy to our plotting of Davis's position. The wind and waves also were moving him along in some hard-to-estimate fashion as time passed. When we finally got to our estimated recovery location, we could see only a few hundred yards, and Davis wasn't there. By then, *John Paul Jones* was about one thousand yards to the west of us.

John Paul Jones had been several miles (four thousand yards) to the west when first ordered by radio to join in the search. Now, as she closed on our position, our two combat information center crews worked out a coordinated search. We had to account for our navigation inaccuracies caused by the difference between the courses and speeds ordered and those made good in the storm's wind and wave flow. Even navigation by radar was difficult because of the ships' pitching, rolling, and the "sea return" on the scopes. *John Paul Jones's* newer design gave her better sea-keeping qualities and much better visibility for bridge watchstanders.

With her in the area, and the hope we were really in the right place, we thought Davis's rescue chances were improving.

Conditions in *Porterfield* had become serious. We made a temporary, below-deck sick bay in the aft living compartment closest to the fantail, where the Oil King and the Hospital Corpsman had been brought by the rescue team. The Corpsman was in shock with a serious leg injury. A depth charge foundation angle iron broke loose and punctured completely through the calf of his left leg. We stopped the bleeding, but his pain was almost unbearable. Even with a healthy Corpsman, we would not have been able to attend to the unconscious Oil King properly. We assumed he had a concussion, but its seriousness was unknown. A young crew member with some ambulance-driving experience took blood pressure and I made the first of several precarious trips from the compartment, up two levels, along the spray-drenched "torpedo deck," to the radio room. There, I relayed what skimpy medical data we had to the ship's doctor in *Midway*. He advised on first aid measures and told us to constantly monitor the Oil King's blood pressure. Then, when we detected the Oil King's blood pressure falling, the doctor advised us he had a serious concussion and gave us no more than three hours to get him to a hospital.

Since *Midway's* helicopters were down, a Navy search and rescue helicopter had been dispatched from Japan to come out and aid the three ships. *Porterfield* was too small for a helicopter landing and too small for a safe highline transfer to *Midway* in rough seas. The safest method for transferring the Oil King and the Hospital Corpsman to *Midway's* hospital, or one ashore, was to hoist them up to the helicopter in a "Stokes stretcher." The inbound helicopter pilot relayed by radio to us that even this wasn't safe in the existing storm conditions. Every choice was a trade between saving and risking lives, with time running out in three hours.

Never had I been called on to make decisions like this. If we abandoned the search for Davis, we could get out of the serious storm effects in about an hour by moving directly into the teeth of the storm toward the east coast of Japan. The mountains of Japan then would start to give us what seamen call a lee, where the wind can't build up the sea as much because of the mountain's protection. That would give us an hour and a half to complete the at-sea transfer and fly them to the Navy Hospital in Yokosuka. *John Paul Jones* could stay and search for Davis and try to transfer him to *Midway* if that was needed after they found him. *John Paul Jones* could make a safer highline transfer at sea than could *Porterfield*. If my weather judgment was wrong, then the search and rescue helicopter might not have enough fuel to last while we worked closer to the mountain's lee. We would lose more precious time while they flew to the *Midway*, refueled, and returned. If *Midway* went with us, we might rob Davis of a desperate need for hospital care. He would have been in the water for about two hours and approaching medically predicted survival limits for the water temperature and wind exposure.

I would not let myself believe Davis might already be gone. After the sacrifice he and the Hospital Corpsman made in trying to reach the Oil King, it seemed unconscionable to leave him. At the same time, I realized his sacrifice was in vain if we let the Oil King slip away.

I decided on my recommendation to the Captain. He was on the bridge, coordinating the search for Davis. I went from the radio room to the combat information center to work out the times, courses, and possible speeds for my recommendation. Now, they could send out the messages as soon as the Captain and the other two ships agreed by radio to the plan. We would leave Davis, *John Paul Jones*, and *Midway* behind, and go for the lee. I climbed the ladder from CIC to the bridge and explained our agonizing choice to the Captain. He agreed with the recommendation and so did the other ships. We changed our course into the teeth of the storm, toward the coast of Japan.

I left the bridge to make another precarious journey along the first level through the sheets of spray. I had to get back to the Oil King and Hospital Corpsman and tell their attending crew members we were getting them to the hospital. Just aft of the bridge, on the port side, is a little alcove to protect the bridge station for the ship's general announcing system. (You see this in many movies where the bosun is piping general quarters over the general announcing system.) I stepped into this alcove and slumped in prayer, for I decided to wait on the daylight-safe course to put up the heavy weather lifelines. They weren't there when three men needed them. And now I was responsible for the decision to reduce Davis's slim chances for rescue by cutting the search effort in half. My prayer wasn't for my comfort, though. It was for Davis and for the crew who needed to understand the choices we had made. I hoped the Lord would deal with me later.

The amen wasn't out of my mouth when the port side bridge lookout called out from the bridge wing that he could see a life jacket in the water. After the next blast of spray passed, I saw it could be a rolled-up life jacket, maybe the one originally attached to one of our lost Stokes stretchers. It "had to be the one" that ripped loose in the same crashing rouge wave that got the Oil King. The Captain decided to pinpoint the search for Davis right where we were, and I called *John Paul Jones*, then about 1,500 yards away. They were heading away from us now and responded they would do a Williamson turn and come down the line of bearing toward us. *John Paul Jones* had not even completed her turn when they spotted Davis less than fifty yards off their bow. If not for that partial turn, they could have steamed past without seeing him in the whitecaps and spray. Just as described in the Pacific *Stars and Stripes* article, he was afloat with the age-old sailor's lifesaver made by tying the ends of his bell bottom trousers, inverting them and pulling them down from above into the water, and trapping air in them. They were as good as water wings.

Davis's discovery took place at least a thousand yards from where providence set in motion my radio telephone call, and possibly even farther from our originally estimated position for him. With joy in our hearts, we told the crew what was happening and plowed ahead for the lee. We made our rendezvous with the rescue helicopter as planned. By that afternoon, our two injured crew members had been treated and were recovering nicely in the Yokosuka Naval Hospital. Davis, in remarkably good health, stayed in *John Paul Jones* for the trip to Yokosuka. When *John Paul Jones* tied up alongside us in Yokosuka, Davis was standing topside, and I walked along our main deck to be as close to him as possible from *Porterfield*. Davis greeted me and showed me a plaque *John Paul Jones* presented him, along with dry clothes. His first comment was, "That was a hell of a way to earn a plaque."

My lesson learned is that major choices occur to all of us in one way or another. Car accidents in "rivers of traffic" or epochal battles in a storm, like that, just related, are part of the unbounded sea of life in which we swim. We make our choices for right and wrong, many of them based on incomplete training and imperfect knowledge of what is around us. Very often we find we cannot control the outcome.

We can call the results luck. We can say coincidence is the lucky or unlucky reason for whatever happens. We can blame ourselves, someone else, the government, or society. Since we are free to make our own choices about whom or what is responsible for tragic or miraculous events, especially in hindsight, we can also look to our Father in Heaven.

Because of Davis's rescue, I believe in miracles. If we turn out to be players in a miracle, we should take no credit. It is like saying our prayers in secret, so they don't become boastful. My real lesson learned is that there is a way to put our faith in action so it may be respected by everyone, regardless of their religious beliefs. If we choose to present Him our plan, ask for His help, and listen to His guidance, the Lord acts to build our character, in humility, while honoring the free agency of everyone else involved. He strengthens us with His love, so we can swim in storms, like those at sea, and have faith, even if the results are not miracles.

Acknowledgments

This book was written by so many people: classmates, wives of classmates, friends of classmates—but more than that, our families and friends who made so many contributions over a year to make this book possible, and the best that it could be. A special *thanks* to the classmates on the Naval Academy Class of 1957 Review Committee* (Bill Baab, Eleanor Boyne, Pete Boyne, Sam Coulbourn, Charlie Hall, Bill Peerenboom, Earl Piper, Bob Strange, and Bob Wellburn) who worked so diligently reviewing and editing our work. Thank you for your help and sacrifices everyone.

Alan Hemphill
Amy Herrin
Beckie Paulk Gowen
Bill Baab*
Bill Hamel
Bill Peerenboom*
Bill Smollen
Bob Christenson
Bob Fox
Bob McElwee
Bob Rosenberg
Bob Strange*
Bob Warters
Bob Wellburn*
Brad Parkinson

Bruce DeMars
Carol Marryott
Carolyn Wisner Andros
Charlie Duke
Charlie Hall*
David Nevin
Don Beatty
Earl Piper*
Earle Smith
Eleanor Boyne*
Fritz Steiner
Fritz Warren
George Bouvet
George Phillips
Gerry Anderson

Jack Homnick
Jennifer Johnsen Beightel
Jerry Barzak
Jerry Dunn
Jim Beatty
Jim Googe
Jim Gradeless
Jim Poole
Joe Koch
John Howland
John Russell
John Stacy
Ken Malley
Larry Bustle
Larry Ingels
Larry Magner
Leo Hyatt
Linda Paulk Reddick
Melanie Johnsen Roper
Mike Giambatistta
Nancy Piper
NASA Official Photographs
New London Submarine Library and Museum
O. C. Baker

Pam Paulk Minor
Paul Roush
Pete Boyne*
Peter Baker
Ray Dove
Rich Enkeboll
Ron Goldstone
Roy Danke
Sabrina Stering
Sam Coulbourn*
Sam Underhill
Ted Almstedt
Ted Kramer
Tom Marnane
Tom Marryott
Tom Ross
Tommy Sawyer
Urb Lamay
USNA Alumni Association and Foundation
US Navy Official Photographs
US Navy Undersea Museum
Walt Meukow
Wilson Whitmire
Zachary Placek

Gallery

Class of 1957 at its 65th reunion in front of the chapel.

Through Bancroft Hall into the Mess Hall for lunch.
The Drum and Bugle Corps provides the marching music.

The Naval Academy Chapel.

The four Iowa-class battleships operated together for the only time in 1954.

The US Naval Academy yard.

322 *Gallery*

A LEAP INTO THE 20TH CENTURY

The N3N, *Yellow Peril*, we learned to fly on Chesapeake Bay.

Second Lieutenant Zach Placek is leaving Bancroft Hall after graduation, as we did, to begin his new career. The photo view of him leaving via the Rotunda was taken by Carolyn Andros from the steps leading to Memorial Hall.

US Naval Academy Chapel.

John Paul Jones's crypt below the chapel.

About James D Paulk Jr.

Born and raised in the small town of Brunswick, Georgia, Jim Paulk graduated from Glynn Academy there, attended the military school, North Georgia College, located in the mountains of Dahlonega, Georgia, and graduated in the Class of 1957 at the United States Naval Academy in Annapolis, Maryland. He served five years on active duty and five years in the Reserve force as a Naval submarine officer. During a business career of twenty-six years with Procter & Gamble, he worked in manufacturing management before starting a business consulting company with other retirees. When he was asked to take on marine conservation projects, his life changed in a new direction. For nine years, he successfully led efforts to eliminate destructive gillnets from California, build artificial reefs, raise white seabass for release into the ocean, and wrote legislation beneficial to recreational anglers. After retiring for a second time, he began writing articles for fishing magazines and newspapers.

He lectured at universities, participated in fishing seminars, did TV roundtable discussions and radio call in programs, and was a member of many fishing clubs including the Tuna Club of Avalon, the oldest saltwater fishing club in the country. Jim was honored with numerous fishing and conservation awards.

After thirty-seven years living in southern California, Jim and his wife, Pat, moved cross country again near Jim's roots in the small town of Kingsland, Georgia located just north of Jacksonville, Florida.

www.ingramcontent.com/pod-product-compliance
Lightning Source LLC
Chambersburg PA
CBHW061744070526
44585CB00025B/2797